# ROOTS
## DEMYSTIFIED

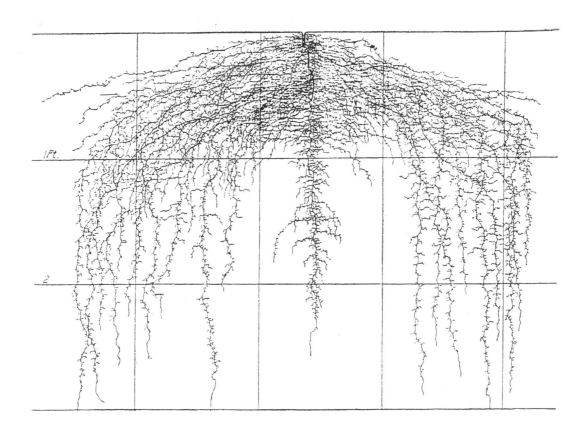

## ...change your gardening habits to help roots thrive

# Robert Kourik

**Metamorphic Press**

Library of Congress Control Number: 2007902280

ISBN: 978-0-9615848-0-1
9780961584832

Distributed in North America by:
Chelsea Green Publishing
P.O. Box 428 85 N. Main St., Suite 120
White River Jct., VT 05001
Orders: 800-639-4099, Phone: 802-295-6300, Fax:802-295-6444
www.chelseagreen.com

Distributed in the UK and Ireland by:

Permanent Publications, The Sustainability Centre

Droxford Road, East Meon, Hampshire GU32 1HR
Tel: 01730 823311, Fax: 01730 823322
www.permaculture.co.uk

Front cover illustration is a view looking down on the
top six inches of the root system of a kidney bean plant.

The title page reveals the cross-section of the roots of a garden pea plant.
(Both grids are in one-foot squares.)

Front and back cover, title page, and template for the page layout designed by Sandy Farkas.

Printed in the United States of America, on 30% recycled paper.

10 9 8 7 6 5 4 3 2 1

Dedicated to:
My constant friend and supportive father, through thick-and-thin

John (Jack) Kourik

# ❧ Acknowledgments ❧

Angie Albini and Virgil Marin (Sebastopol, CA) Both work at Harmony Farm Supply and provided valuable assistance in making sure Figure #70 was accurate.

Lynda (not misspelled!) Banks (Novato, CA) thanks for proving the attractive, overall design for my Web site (www.Robert-Kourik.com).

Kathleen Barber (Underwood, MN) produced the index with a keen eye to how a reader accesses a book's information.

J. Renee Brooks (Western Ecology Division, U.S. EPA/NHEERL, Western Ecology Division, Corvallis, OR) helped provide papers and advice about hydraulic lifting in Douglas fir (*Pseudotsuga menziesii*) forests.

Dr. Efren Cazares (Corvalis, OR) for his insightful review of the chapter: "The Good Fungus Among Us."

Todd Dawson (Dawson Lab, Life Sciences Building, UC Berkeley, CA) was kind enough to send me dozens of papers about hydraulic lifting and hydraulic redistribution, and review my attempts at making this fascinating topic understandable.

Sandy Farkas (Forestville, CA) provided expert graphics assistance: designing the template for the look and type faces for each page, the design of the cover, and 20 of the pen-and-ink drawings.

Dr. Charles R. Hall (Professor, University of Tennessee Extension, Dept. of Agricultural Economics, Knoxville, TN) for the data on the amount of time it takes to mow a lawn.

Dennis Hansen (Sausalito, CA) provided valuable assistance in the details of subsurface drip irrigation.

Sherry Havens (Santa Rosa, CA) for her loving support during the production of this book.

Anne Hiaring, Esq. (San Anselmo, CA) who provided simple-to-understand legal advice.

Amie Hill (Graton, CA) has been my steadfast editor over the years. She has added wit, style and correct grammar to this book and many others.

Patty Holden (Sebastopol, CA) is the computer wizard who helps Sandy and me through the tortuous world of InDesign computer graphics.

Greg Jorgenson CIT, UC (Fresno, CA) helped provide data about the efficiency of subsurface drip irrigation and for telling me that after a number of years those pesky gophers finally found and ate through his tubing.

Robert Kourik (Occidental, CA) did all the layout based on the template provided by Sandy Farkas. Any mistakes or odd layouts are my responsibility and not a reflection of the excellent work of Patty Holden and Sandy Farkas.

Cornelia Krause (Sciences Fundamentals, Université du Quebec a Chicoutimi, Canada) for providing the research and illustrations of how to plant young seedling trees.

Wendy Krupnick (Santa Rosa, CA) provided both much appreciated moral support and technical advice.

Frederique Lavoipierre (Sebastopol, CA) and other Sonoma State University reviewers helped review the complex and interesting workings of how roots grow. And what exactly is mycorrhizal association anyway? She helped explain.

Marshia (not misspelled!) Loar (Occidental, CA) thanks for providing a beautiful, serene and quiet place to live and work.

Richard Merrill (Scotts Valley, CA) helped edit the chapters "How Roots Grow" and "Humus & Mulch."

Suzanne Nelson Ph.D. (Director of Conservation, Native Seeds/SEARCH, Tucson, AZ) helped estimate how far apart the Hopi Indians planted their corn.

Marty Roberts (Sebastopol, CA). Thanks for helping me organize, build, and manage my Web site. (www.Robert-Kourik.com)

Laine Velinsky (Novato, CA) guided me through the ups-and-downs of this self-publishing adventure.

Michelle Vesser (Occidental, CA) and Richard Molinar (Farm Advisor, Small Farm/Specialty Crops, University of California Cooperative Extension, Fresno, CA) for help unraveling the confusion about symphylans.

Margie Wilson (Graton, CA) looked over the manuscript with an eagle-eye's attention to detail of both the words and all the illustrations' layout. Any typos are because I overlooked her notations.

# ❧ Table of Contents ☙

# Introduction

**R**oots. When they're part of your family tree, you cherish them. In the context of plants, however, to paraphrase the cantankerous comedian Rodney Dangerfield, they don't get no respect. This could be because they're frequently ugly, mostly invisible, and practically nobody understands them. Even deprived of the appreciation they deserve, plant roots perform some of the most vital functions on our planet, providing, in myriad forms, sustenance and support for most plants and, by extension, for much of our human life and activity.

As a fledgling landscaper in the 1970s, I began to get curious about how tree roots actually grow. It wasn't long before I was hooked on collecting examples of excavated root systems from bulldozed orchards and keeping an eye on roadside cuts to spot the fascinating twists of wood left behind there by deep soil erosion. In many subsequent years of mucking around in all manner of soils and surroundings, I dug up a lot of roots. Even as I write, prize examples from my extensive, eccentric and instructive root collection hang naked and marionette-like from tree limbs near my house, simultaneously illustrating and exposing truths and myths about this hidden and essential plant part.

In the early years of my landscaping career, however, my powers of observation and intuition must have been stymied by the then-common belief, frequently repeated by horticultural "authorities" (who, I figured, knew much more than I), that the roots of a plant extended only as far and wide underground as its foliage.

Then, one day in the early 1980s, as I was reading a popular gardening book, I suddenly noticed that the book's illustrations of trees consistently showed each specimen's roots as an upside-down mirror of its canopy. My dormant intuition suddenly started flashing warning signals. What? A Christmas tree has deep pointy roots? An oak has a rounded root system like its top? Something just didn't seem right in that orderly little world.

As a result of my gut feeling, I began to rummage around in agricultural libraries (there was no Internet then) in the hope of finding photos and drawings that showed actual excavations of actual root systems. What I discovered radically changed how I looked at roots. I learned, for instance, that the area occupied underground by tree roots can be up to five times more, or greater, than that of the foliage above ground, and that, frequently, one-half or more of a plant's mass is located below the surface of the soil.

**T**hat's what this book is all about: the facts, not the myths. Here are just a few examples of what I discovered about roots and their hidden and marvelous activities:

- Some roots pass water on to the nearby roots of other plants.

- Feeding roots often tend to grow up rather than down.

- At the end of its first year's growth, an apple tree can produce as many as 17,000,000 root hairs with a total length of well over a mile!

- Roots can exude chemicals that dissolve minerals.

- Many trees do not have taproots (single roots that grow deeply into the earth).

- While feeding near the surface, some shrubs, like *Artemisia tridentata*, can send their roots as deeply as 30 feet.

- Some trees (*Juniperus monosperma*, for example) have been found with roots growing 200 feet deep.

- About 90% of a tree's roots are to be found in the top 18 inches of the soil beneath and around it.

- The common alfalfa, while not a very tall or imposing plant aboveground, can vary its root depth from slightly over one foot to 128 feet, depending on the soil.

- A measly turnip can produce roots that explore 100 cubic feet of soil (enough "dirt" to fill 20-25 wheelbarrows), and the roots of the lowly lima-bean bush as much as 200 to 225 cubic feet.

- A sprouting cucumber seed, in a good, loose topsoil, can grow a taproot down to about three feet, at the rate of one inch per day.

- The roots of a single rye plant can extend for a total of 372 miles and have 6,123 miles of roothairs.

These are all fascinating facts, but what do they mean to a gardener? That's the thrust of this book—what can studying roots teach us about cultivating and nurturing our own gardens and landscapes? In these pages, in addition to showing you the root-growth patterns of each plant I discuss, I'll investigate the implications of these patterns and describe how this knowledge can change your own gardening behavior in terms of protecting roots, nurturing the soil, ensuring well-placed soil fertility measures, and the wise use of water and mulch.

In the chapter "How Roots Grow," there's a bit of biology about the nature and construction of roots and root hairs. (Depending on your own nature, you may find this section fascinating or boring, but it's ultimately extremely useful.) Then, I'll pass along some stories about our good friends "Humus and Mulch," and show how the soil that supports and feeds roots is dependent upon and enhanced by a healthy amount of these invaluable substances. Next comes a section on "The Lawn"—the largest areas of root systems in most of suburbia and many cities—followed by an exploration of "Prairie Grass" and "Shrubs."

Oh, and you'll be pleased to discover that the following chapters get even juicier as we root around in the subjects of edible and fruit-bearing plants.

Enjoy!

Robert Kourik
Occidental, CA

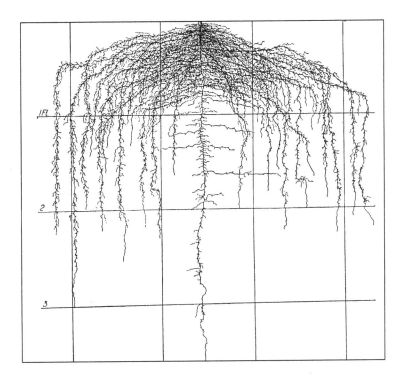

Can you guess what plant has this root system?
(Each square is one-foot square.)

# CHAPTER 1

# How Roots Grow

[NOTE: The chapter that follows is fairly technical and fact-filled. In addition to providing necessary down-to-earth introductory information, it also functions as a solid source to consult when the intriguing chapters that appear further on whet your interest for detailed information.]

Next time you're outside, take a good look at the plants around you. Depending on where you live, you may see trees, shrubs, lawns, field crops, cactus, gorgeous gardens, beach-side grasses, prairie lands—all part of the vast aboveground range of plant life. Then take a moment to consider that most plants you see have a shadow self reaching underground to nourish and sustain it, an entirely invisible world of growth, a massive eco-drama going on under your feet—the complex and mysterious world of roots.

According to eminent Russian microbiologist N. A. Krasilnikov, a single two-foot-tall alfalfa plant can produce up to 14.3 billion root hairs. Impressive as this statistic may sound, it's just an example of business as usual in the opportunistic underground world of roots, which invariably move to their own best advantage, creating curious and changing patterns in the soil. Roots and their root hairs are also unpredictable; a single species may differ wildly in form from plant to plant and in varying soil conditions, while faithfully preserving the same basic qualities, inclinations, and mechanisms.

So how and why do roots do what they do? How do they affect our soil and water and air, and, for that matter, most of the planet? And what instincts and processes drive these vital lifelines inexorably through the earth, even against the greatest of odds?

## Birth of a Root

A root begins in the magic moment when a seed sprouts. The scientific description of this miraculous process is simple: the *apical meristem* (also called the *primary meristem*) that is present in the embryo (seed) of a plant, stimulated by outside conditions of temperature, moisture, and soil chemistry, begins to grow. The meristem of a seed is (1) a general zone where undifferentiated cells frequently divide, and (2) the locus for a group of actively dividing embryonic cells that are specialized for the production of specific types of new cells. [Refer to Figure #1 for a visual representation.] The leading growth cells of a root's meristem are known as the apical meristem because they're located at the apex (tip) of a root.

At the tip of each potential root is a thimble-shaped mass of cells, called the *root cap*, which covers the apical meristem. The root cap consists of a loose grouping of cells held within a slimy substance called *mucigel* (a pectin exuded during the growth process) that protects and lubricates the root tip as it finds its way through the soil.

An aboveground shoot, by the way, has an apical (tip) bud, much like a root's apical meristem but without the root hairs or a root cap. Aboveground shoots differ from roots in that their cells are specialized to photosynthesize. Photosynthesis is a plant's way of combining

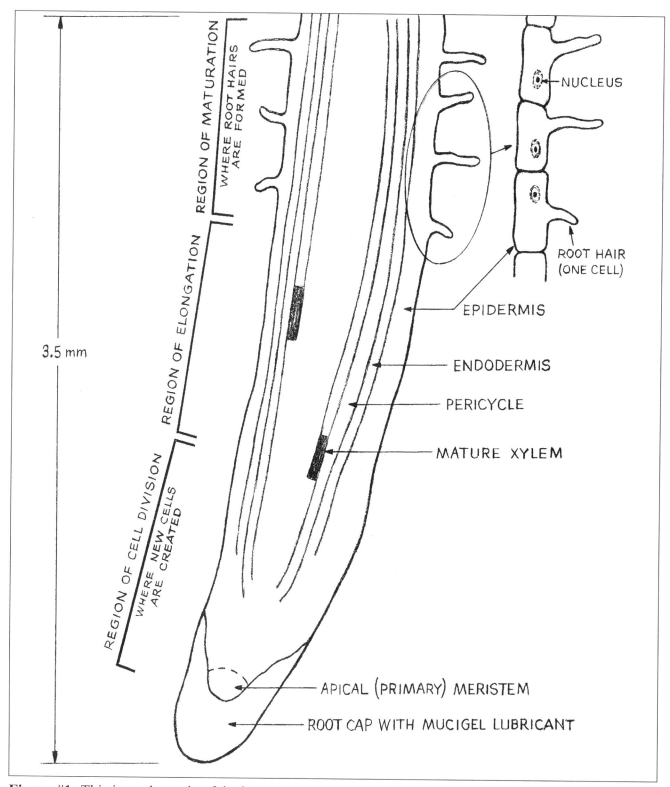

**Figure #1:** This is a schematic of the important parts of a growing root and root hairs. There are three important "regions" to a young root—cell differentiation, cell elongation, and the region of maturation. The root hairs, which absorb nutrients and usually live for only one day, are found exclusively in the zone of maturation. Some root hairs do live to become white roots on which more root hairs can form.

carbon dioxide with the energy of light and water to make sugars (glucose). Fortunately for us, this chemical reaction produces a "waste" product—oxygen. The photosynthesis-produced sugars in stem tissues fuel the amount of energy required for movement (called *active transport*) of nutrients into the cells of root hairs. All this occurs at the molecular level.

Another example of the division between root and stem tissues is the green "shoulders" of carrots. These are actually stem tissue, since root tissue can't photosynthesize. The bulbs or tubers of crops such as garlic, onions, and potatoes, as well as ornamentals like daffodil, freesia, and bearded iris are actually modified underground stems.

Behind the protective root cap and the apical meristem (known as the *region of cell division*) is an area of cell differentiation known as the *region of cell elongation*, followed by a section known as the *region of maturation* in which root hairs are formed. The region of maturation (see illustration) is the only part of the roots where root hairs are found. Each root hair is composed of only a single epidermal cell, which is usually very short-lived and functions for only a short while. (The productive life of most root hairs is about one day, which is why new ones are always forming.)

Since they are the source of all nutrient exchange between the plant and its surrounding soil, root hairs have an enormous surface area for their size. And since each root hair's lifespan is so short, one goal of the gardener is to develop plants with a healthy *lateral* root system (in other words, one in which the roots grow horizontally or obliquely from the "woody" downward-reaching portions of the root system). Encouraging lateral root growth provides large areas in which root hairs can arise, since they will only form on growing tissue, not on solidified, woody root material. [Figure #16 in the discussion of cabbage and cauliflower roots describes how to encourage lateral roots when growing vegetables. In the case of trees the taproot is usually eliminated upon planting, leading to more laterals. (See illustrations #9 and #35.) Lateral growth can also be stimulated with judicious root pruning at planting time.]

Once formed, root hair cells, which need moisture in order to stay alive, obtain it via the process of osmosis. *Osmotic pressure* occurs when a liquid (in this case, water molecules), passes through a selectively permeable membrane, such as a cell wall. In root hairs, this happens when the moisture content within the cells is lower than that of the surrounding soil moisture. Water obtained by the root hairs is also required for the process of photosynthesis. Because there are more nutrients inside the root-hair cells than outside of the cell membrane, in order for nutrients—salts and sugars—outside the cells to enter the root hairs, the cells must expend energy to absorb the nutrients against osmotic pressure.

## Roots of a Different Color

Xylem are the long tube-like plant cells that conduct water from the root hairs (via the lateral roots) up the length of a tree, shrub, or perennial plant. The root xylem form near the zone of elongation, behind the region of cell division. There are two very important layers of cells surrounding the xylem; the outer layer, called the *endodermis*, forms a thick corky barrier that forces water and nutrients drawn into the xylem to stay there and prevents leakage out of the lateral root. Inside the endodermis is the *pericycle*, which still has the capacity for new growth by cell division. The pericycle is where new branching lateral roots originate and where thickening will occur if the lateral root survives and becomes woody.

In woody plants, the functioning lateral roots that generate root hairs for absorbing nutrients and water come in two forms: *woody roots* and *white roots*. The woody lateral roots function as a kind of "underground trunk" for a plant and are the roots that form the immediately visible root system of a perennial plant, shrub, or tree. White roots, on the other hand, act as "sites" for the active growth of new root hairs. The white roots are visible to the naked eye at the ends of the brown, woody roots. As a white root ages, its cortex dies and its endodermis (the barrier layer) is exposed to the soil. At this point, the white root turns brown and sometimes dies. Not all brown-colored roots are dead, however. If the root lives (a random process), new roots from the pericycle may burst out through the old covering. Occasionally, a formerly white root lives and thickens to become part of the woody lateral root system that anchors the plant.

## Maturing Roots

Where trees and shrubs are concerned, the first root to form when a seed sprouts is known as the *primary root*. From this primary root grow the secondary roots, or laterals. When the primary root grows more rapidly than the secondary lateral roots, a *taproot* is formed, growing straight down into the earth instead of outwards. Botanically speaking, *fibrous roots* are those that arise from the stem tissues of monocots. Examples of monocots (monocot is Latin for "single leaf") include wheat, rye, palms, lilies, and all grasses.

When the taproot of a woody plant (or perennial) grows slowly or ceases its growth,

and the numerous secondary roots take over, a *fibrous* root system results. Such a system is characterized by the production of many major horizontal roots (laterals) and oblique roots (growing downward at an angle from the surface of the soil), all originating at the base of the trunk (also known as the *crown* of the root system). Many trees, including most fruit trees, have fibrous roots. [See Figure #2.] Many of the larger roots tend to grow downward (there are some fascinating exceptions, as we'll see in the chapters on native shrubs and trees); if the lateral roots don't grow very far down, due to obstructions such a clay or hardpan layer, the root length may develop extensively near the surface.

Most of a plant's horizontal roots grow within the top one to three feet of the soil (even most trees), and there can also be numerous vertical roots, called *sinkers*, which may descend anywhere along the length of the laterals.

The death of a root is like a deep composting process. The various woody parts of the root that have died are devoured (and their available nutrients "liberated") by the many types of microbes found in the soil.

The metabolism of microorganisms (as well as many chemical and biochemical processes, and the transformations of various organic and mineral substances) is more intense in the root's *rhizosphere* region, an area of microbial activity in interaction with root *exudates* (see below), which immediately surrounds each root hair. In the mysterious region of the rhizosphere, various minerals, rocks, limestone, marble, and other raw materials are decomposed at a faster rate. This process is not only caused by root excretions such as carbonic acid [carbon dioxide ($CO_2$) plus water ($H_2O$) equals $H_2CO_3$ or carbonic acid] and other acids, but also by the microflora of the rhizosphere. The more intense the growth of

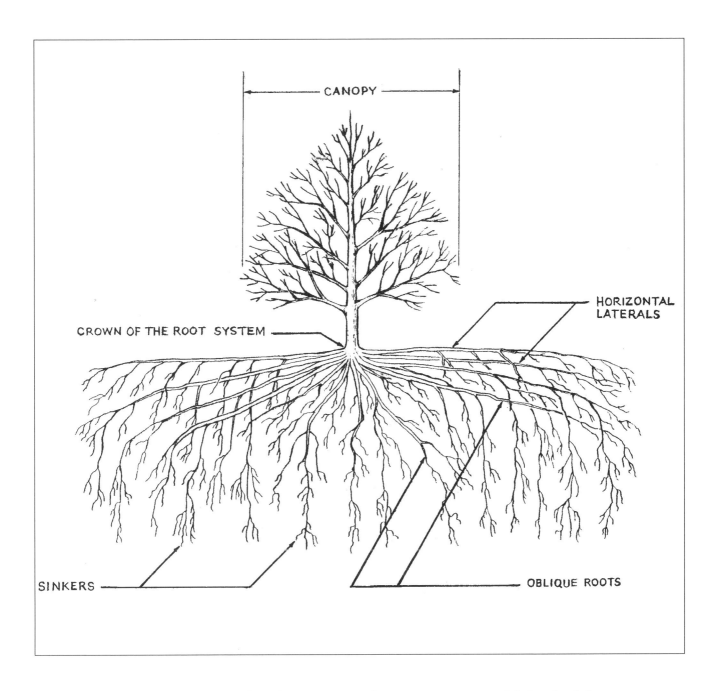

**Figure #2:** True fibrous roots are found on monocots (grasses are an example) that don't have a taproot. Fibrous roots originate at the stem (base) of the grass. Many trees start to grow with a taproot, but change after a few years to a well-spread fibrous root system with horizontal, oblique, and sinker roots for stability. In the case of trees, the horizontal, oblique, and sinker roots are called *fibrous roots*.

microbes, the faster the decomposition process of the substances that come within reach of them. Certain compounds, for instance, tricalcium phosphate, are made available by the soil microbes found in the rhizosphere. [For more on the rhizosphere, see the next chapter.]

One of the tasks of horticultural microbiology is the enrichment of the root-hair region with microbes, which help to transform non-soluble phosphorus and other nutrient compounds into the soluble compounds available to the plant. A good gardener will focus on the enrichment of the root's rhizosphere with the microbes found in compost and green manures (fertility obtained by turning under green foliage).

Decayed roots also provide channels for earthworms to travel and convert woody tissue and other organic matter into highly nutritious worm castings (feces). Likewise, plant and tree roots take advantage of the wormholes as easy places to expand their growth.

When it comes to moisture, roots are lazy. They won't grow *to* a water source, but will grow *where* there is moisture. Roots in dry soil show very little growth, except in plants specially adapted to desert regions. At times when the soil is moist and the surface is warm, roots near the surface show the most growth. If the soil dries out, lateral roots near the surface will die and new roots will grow more deeply into the soil. As a dry season progresses, the depth of active root growth moves progressively further and further into the soil. With each rain or irrigation, new lateral roots are produced near the surface. These new lateral roots absorb water and nutrients, but many will die as the soil dries out again, or as they are replaced by new feeding lateral roots.

## Active Roots

When it comes to nutrients, however, roots don't just sit around waiting for some good stuff to show up; they actively "mine" the soil, producing a wide range of exudates (chemical materials that ooze out of the root hairs). Sometimes aptly called "the cake and cookies for soil bacteria and beneficial fungi," the exudates also make previously unavailable nutrients more useful to the plant, with the uptake of phosphorus being one frequently cited example.

The presence of exudates was observed for the first time in the middle of the 19th century. It was established that, when in the presence of acidic compounds in the root excretions of plants such as barley, wheat, oats, foxtail, and other similar crops, lupine plants excreted substances of an acid nature that dissolved highly soluble phosphates in their surrounding soil, transforming them into an easily assimilated form. It was observed that other plants that are not able to excrete substances of an acid nature cannot dissolve mineral compounds.

Now that we've covered the more technical aspects of root growth, we can now move on to the tasty subject of compost, the backbone of any healthy garden or orchard. Roots love compost, sheet composting (layers of fresh green and dried plant and manures spread over the surface of the ground, like a two-dimensional compost pile), and mulch. Compost in particular is a fast way to enhance the populations of beneficial soil flora and fauna, but I'm not going to drone on about how to make it. Rodale Press alone has probably published more books on the subject than you can shake a worm at. I'll leave you to wade through the many articles and books peddling their version of the "correct" way to make compost. What I will stress is that compost improves the tilth (consistency) of the soil and raises its humus content…which leads us to the next chapter.

FOOTNOTE

Here are some more examples of excretions.

- Plant roots excrete organic compounds (also called exudates) such as acetic, formic, and oxalic acids, as well as organic acids, sugars, aldehydes, ethyl alcohol, and other compounds.

- Acidic compounds are found in the root excretions of lupine, peas, buckwheat, mustard, and rape (canola). The amount of phosphorus that was made available by these excretions was 14–34 % of all the phosphoric acid absorbed by the plant.

- When detected organic substances were found in the root excretions of lupine, beans, corn, barley, oats, and buckwheat, the excretions reached their maximum during the fourth week of growth (although at a somewhat earlier period in buckwheat). Upon the ripening and aging of the plants, the amount of root excretions decreases and toward the end of the growth period stops altogether.

- Nitrogenous and non-nitrogenous organic compounds can be found in the root excretions of corn. The amount of nitrogenous substances in root excretions decreases with the age of the plant.

- More recent studies are looking at the possible use of plant exudates (mostly plants other than trees) for phytoremediation—to remove or neutralize polluted soil or water. These projects target heavy metals, metalloids, petroleum, hydrocarbons, pesticides, explosives, chlorinated solvents, and industrial by-products. Removal or degradation of these compounds is thought to occur in the rhizosphere—the region of the soil influenced by the presence of plant roots, microbes, and other soil fauna.

FOOTNOTE

# CHAPTER 2

# Humus & Mulch

Here's a quick description on the living dynamics of healthy soil, and a close look at two aspects of soil composition that affect every plant in your garden—*humus* and *pore space*.

First, some useful definitions to dig into. The good news here is that while texture is predetermined, any gardener can improve soil structure, either quickly or slowly, by using techniques such as careful spading and cultivation, composting, cover cropping, green manuring, mulching, and sheet composting (more on these later).

• The *texture* of a soil refers to the size of its mineral particles. Texture is determined over geological time, and can range from gritty sand to pasty, silky clay. It is the relative proportion of sand to silt to clay. Since a soil's basic texture is the result of eons of natural, geological progression, gardeners are pretty much stuck with it. Another way to approach the definition of texture is the relative surface area. Here's a staggering comparison measurement: it takes 65 million clay particles to fill up the same amount of space as one grain of sand. Thus, clay has a much higher surface-to-volume ratio.

Mineral particles (sand, silt, and clay) are shaped differently and arrange themselves differently when grouped together, producing soil variations such as loose and sandy, crumbly and light (loamy), compacted into heavy clay, or anywhere in between. This is because of their textural difference. Clay, for instance, is in the shape of flat crystals (also known as "plates" or "flakes"), which can pack together tightly and squeeze out air, resulting in a dense, compact, slick soil.

Simply put, over a period of geological time rocks weather into sand, then into silt and finally clay—texture. The gardener cannot reproduce this slow process of nature.

• But while you're pretty much stuck with soil texture, you *can* change the soil *structure*. Structure refers to the way the mineral particles (sand, silt, and clay) are arranged into *aggregates* (groups of particles that are loosely held together). Examples include small crumbs, small to large blocks, tall skinny columns, etc. (mostly viewed on a microscopic level, although you can sometimes see small pea-sized aggregates on the soil surface). Soil structure is a by-product of the decay of organic material. The structure is formed as small clods, depending on the way the microbial life of the soil binds together the three textural elements—sand, silt, and clay. An old guideline goes: "Change structure, not texture, and do it with organic matter."

Organic matter is biologically converted into compost, which then breaks down into humus—structure.

## Good Humus, Pore Space

*Humus* (pronounced "hyoo-mus," and not to be confused with the popular Mediterranean dish made with garbanzo beans) is the end product of organic decay which is constantly being formed beneath a surface layer of *duff* (plant material that is still recognizable and not yet rotted or decomposed) by the action of soil organisms and other conditions. In *The Nature and Properties of Soils*, Buckman and Brady (page 142) define humus as "...a complex and rather resistant mixture of brown or dark brown amorphous and colloidal substances that have been modified from the original tissues or have been synthesized by various soil organisms." How's that for scientific lingo?

Take a patch of uncultivated or mulched ground and scrape off the surface duff. Just below this undigested litter, you'll find a dark layer of decomposed plant matter. This is some of the humus, created primarily by microbes and other soil fauna—aerobic bacteria, microbes, fungi, beetles, etc.—as they digest and excrete "raw" materials (fallen leaves, grass clippings, compost, etc.). With all these elements living, eating, being eaten, excreting, rotting and dying, humus is constantly being formed into a fairly stable and complex compound that serves to hold the pore structure open. (Yet, it's unstable enough that it constantly needs to be replenished, or else we would be up to "up to our necks" in humus.)

## Pore Space

The terms *"pore space"* and/or *"pore structure"* refer to the maze of minute continuous channels found throughout the upper layers of most soils. Organic matter improves pore space by helping make an aggregate of mineral particles form more continuous channels.

Richard Merrill, Professor Emeritus, Department of Horticulture, who taught organic gardening and farming classes for 32 years at Cabrillo Community College in Aptos, California, puts it this way: "...clay has more pore space but drains slower because the pore spaces are large and not continuous with one another like they are in sand. The analogy I use is a bucket full of golf balls and one full of head pins. The former has less pore space but drains better because the pore spaces are large and continuous, not more abundant. The main advantage of improving structure is that you can make a clay soil drain well by [aggregating] the clay particles into grainy or crumb structures...under the influence of the by-products of organic decay." (Merrill, personal communication.)

An ideal mineral/humus/pore structure balance results in a crumbly soil that allows water to percolate down, harmful gases to vent, and refreshing air to permeate the soil. Soil breathes 24/7 at a lumbering, beneficial rate we cannot hear. A soil with a healthy structure allows for easy and deep root growth and will produce the best-looking lawn, garden, and tree growth.

So, here's another of life's marvelous cycles: in ideal conditions, nature continually renews organic matter with plant growth, which in turn partially decomposes into humus. Now another vital ingredient comes into play: the biological action of certain organisms in the humus coats the surfaces of the soil's minute particles to form a colloidal matrix—a kind of gooey, slimy chain of molecules called polysaccharides. This coating helps to maintain a thin film of beneficial moisture from rain or irrigation and assists in keeping the pore spaces open, thus creating an ideal medium for plants to flourish.

Most soils will contain more humus near the ground surface, since the highest population of the soil organisms that decompose raw fiber into humus tend to hang out in the most aerobic zone of the soil, that is, in the duff or just under it. [See Figure #3.] The upper horizon of the soil is also the place where the most nutrients are liberated. Soil flora and fauna act as nature's little fertilizing machines, using the creation of humus, among other processes, to liberate unavailable nutrients into a soluble form that can be absorbed by tiny root hairs—a process known as "mineralization."

One study estimates the number of the bacteria in a gram of soil taken from upper layers of soil surfaces as ranging from 58 million to as many as 3–4 billion. Dig and test just three feet lower, and the bacteria numbers drop to as few as 37,000 per gram.

However, it's not just the soil's humus-clay-moisture complex that liberates nutrients. As mentioned earlier, plant roots, stimulated by the action of organisms in the humus, aid in the nutrient-release process by exuding sugars, organic acids, and other compounds to stimulate the microbial action in the rhizosphere. Thus, roots and humus essentially function as partners in plant growing.

Plants primarily absorb most of their nutrients in a chemical process called "ion exchange." This is a process in which ions (an atom or a group of atoms that has acquired a net electric charge by gaining or losing one or more electrons) are exchanged between a solution and an ion exchanger, i.e., an insoluble solid. Two notable ion exchangers are clay and humus, which are found suspended in the thin, moist

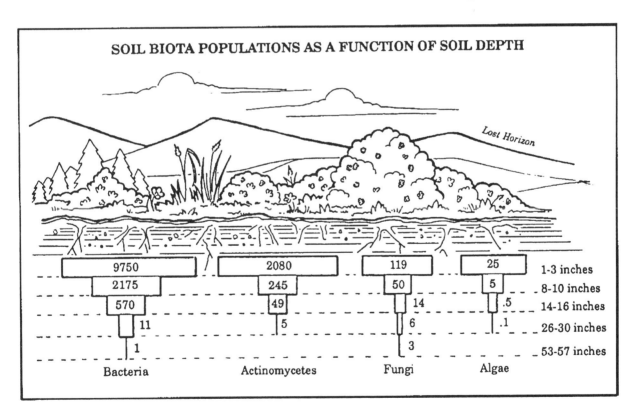

SOIL BIOTA POPULATIONS AS A FUNCTION OF SOIL DEPTH

| | | | | |
|---|---|---|---|---|
| 9750 | 2080 | 119 | 25 | 1-3 inches |
| 2175 | 245 | 50 | 5 | 8-10 inches |
| 570 | 49 | 14 | .5 | 14-16 inches |
| 11 | 5 | 6 | .1 | 26-30 inches |
| 1 | | 3 | | 53-57 inches |
| Bacteria | Actinomycetes | Fungi | Algae | |

**Figure #3:** This illustration shows how dramatic the difference is between the surface-loving soil life and soil life just a bit deeper. Tillage disrupts this natural layering until the various "crittters" have a chance to repopulate the level of soil they prefer the most.

coating of the soil's structural aggregates. The chemical and biological activity in this thin layer of moisture converts nutrients into a soluble form that roots can absorb via ion exchange. Humus binds the clay particles so that the clay forms the aggregates that help maintain a more continuous pore space. The perplexing nature of a healthy humus-clay structure is that it both holds onto and releases many of the nutrients plants utilize.

Clinging to the soil's clay and humus is a thin film of moisture. Too much water produces an unhealthy, anaerobic soil. Too much air in the pore space means that root hairs shrivel, die, and oxidize, as if in the slow burning of an invisible fire.

Lack of pore space due to a deficiency of soil aggregates and enough humus to form the desired aggregates are factors which result in heavy clay soils.

Sandy soil, on the other hand, contains far less humus due to its lack of organic matter and, with its excessive amount of physical pore space, can't hold onto moisture and nutrients efficiently. Also, any humus that might try to form in a sandy soil is oxidized by the high amount of oxygen, which permeates the sand.

Loam soils are created from the ideal mixture of sand, silt, and clay. Loams also maintain reasonable drainage, due, in part, to their humus content—the larger pore spaces allow for the penetration of water and the exchange of gases, while the smaller pores retain moisture. The wonderful structure of a crumbly, loamy soil is due to a healthy amount of pore space held together by a complex web of soil particles and colloidal humus in various sizes of aggregates. (The Holy Grail of an ideal loam is 15% clay, 40% silt—particles that are irregular in shape, fragmented in shapes unlike the plates of clay—

and 45% sand.) Good soil is a natural balancing act.

Don't walk on or till soil that is too wet. I once advised some garden interns that the soil they were double-digging was too wet. Still, they proceeded to form a bed with soil raised 2–6 inches above the path. By midsummer, the clay structure had simply collapsed, and the surface of the bed was *below* the level of the pathway! They had created the temporary look of a mounded, raised bed by "fluffing up" the soil, but the action of the shovels actually compressed what little pore (air-holding) space was already in the clay. As the soil settled, it sank below path level because the weight of the compressed clay squeezed out the air between its plates and caused it to become even more anaerobic and flat. (Potters knead and work their clay on purpose to exclude air and make the clay more dense and smooth.)

## ❧ PRACTICAL TIPS FOR GARDENERS

I have three words of advice for the home gardener—compost, compost, and compost. *Up to a point.* The more organic matter you can incorporate into your soil, the more likely you'll maintain a healthy organic matter content of two to eight percent. A good range for sandy soils would be 2–4%. Loams at 3–6%. And, clay should be around 5–8%. A soil test from a reputable horticultural lab will reveal your garden's organic matter level.

Be sure the compost you apply to your garden is thoroughly decomposed. A "finished," properly-aged compost is no longer hot and makes no "steam" when turned or off-loaded from a commercial supplier. The finished material should have almost no recognizable pieces of the original compostable matter. It should also have the sweet smell of a forest loam. To achieve the goal of finished compost, you need to turn the pile two or three times (maybe even more) to incorporate oxygen into all the raw, composting materials. Then let the pile age, so that it develops a large cross section of microbes and other beneficial soil flora and fauna. The process may take up to one year, so plan in advance and always have a pile going for the following garden season. (Using worms to compost kitchen scraps is like a fast compost bin. The raw materials are quickly converted to castings—manure— that both stabilizes and innoculates organic matter better than unfinished compost.)

Commercial compost is often turned, but because of the surface area required by the large quantities made to meet high commercial demand, it is usually not cost-effective to both properly turn *and* age the compost. Thus, commercial compost is often sold in an unfinished state; beware if you have a load delivered, and it is still hot and steamy. Such compost should be allow to "mellow" until it has a dark, loamy feel, and it may require more turning.

If too much unfinished compost or fresh manure is added in great quantities, you risk the scourge of *symphylans*—nasty little critters that thrive in sandy loam soil, soils with a high level of organic matter and friable (crumbly) soil. Symphylans are 1/4-inch long and look like white centipedes. They eat the roots of many vegetable plants and are nearly impossible to banish by any organic method(s). However, a fallow period may be one option. Compulsive "Captains of Compost" are seeing more and more of this horrible pest. It can now be found in the northeastern, north central, and western United States. Beware of applying too much unsifted compost to your garden, especially in sandy loams. The addition of a layer of more than one to two inches may be too much. Keep the organic matter between three and five percent. Check with a lab report.

Maintaining fixed paths between beds keeps the soil from needless compaction and allows for the growth of the delicate mycelium of beneficial mushrooms. [See Mycorrhizae as discussed in the "The Good Fungus Among Us" Chapter.] Plan the width of vegetable and flower beds so that all plants are within easy reach from a path. I've found that keeping the width of raised beds to three feet or less will protect both the soil and the lower back muscles—especially important for aging baby-boomers.

## I like to mulch a lot, mulch a lot, mulch a lot.

Metaphorically speaking, roots and mulch go hand in hand. The pattern of root growth influences how a gardener mulches and vice versa. Ruth Stout, the diva of no-till gardening, says she didn't invent deep mulching; rather "God invented it simply by deciding to have the leaves fall off the trees once a year." [Ed. note: Even conifers and "non-deciduous" trees will form a good layer of humus in the ground beneath them; the needles and leaves fall over much of the year, and are cast off in somewhat larger quantities in the autumn.] As Ms. Stout describes in *How to Have a Green Thumb Without an Aching Back*, deep mulching can be used as an alternative to tillage by layering compostable materials over the garden surface (sheet composting), followed by mulching. This builds the soil's structure from the top down by mimicking the natural decomposition common to forests and fields.

Ruth Stout devoutly followed a regimen of applying deep layers of spoiled hay (rained on and too moldy to use as feed) without turning the soil—even with asparagus. She maintained good yields, while suffering few pest problems, except for Japanese beetles. (Since the grub of the beetle lives in the ground, often in lawns, she and her neighbors treated all their lawns around the mid-1950s with DDT. Yikes! Not what Rachel Carson would recommend!) A more environmentally sound way is to set up traps to catch the flying adult beetles and apply milky spore disease and/or predatory nematodes (*Heterorhabditis bacteriophora*) to the lawn.

It must also be noted that Ruth had been gardening on the same spot and had added numerous loads of manure for 14 years before starting her deep mulch/no-till garden. So, no-till was preceded by lots of tillage, a fair compromise.

I should mention that throughout this book the word "mulch" will appear to have a multitude of different meanings and different uses. Some mulches, including plastic (polyethylene) sheets, landscape fabrics, sand, large-sized wood chips, and ornamental rocks, are used solely to discourage weeds and/or conserve moisture. Most mulch materials, however, are meant not only to conserve moisture and discourage weeds, but also to improve the soil. The list of these nutritional and texturizing mulches is extensive and includes small-sized chipped bark, shredded tree trimmings, buckwheat and/or rice hulls, cocoa bean shells, compost, shredded cornstalks, cottonseed hulls, grass clippings, washed cow manure (the solids left after water is pumped into holding lagoons at dairy farms), leaves, leaf mold, grape pumice, newspaper, rolls of horticultural paper, hay, and straw. As you see, the list of potential mulches is only limited by a gardener's creativity and the resources available.

All mulches have limitations and unique benefits, and the question of which to use will be determined by your climate, soil, planting situation and other specific needs. For instance, I live on the edge of wine country in Northern California, and I can easily get lots of grape pumice (the seeds, stems, and skins of wine grapes after they've been pressed). Grape pumice is on the acidic side, which makes it a good mulch for blueberry bushes and strawberry plants, but it is fairly high in woody material and takes a longer time to break down when tilled under. Repeated application of tilled-in grape pumice has worked well, however, for one garlic-raiser I know. For him, it's just a matter of waiting a few weeks or more before planting, so that the nitrogen in the pumice is no longer tied up by the increased population of soil microbes that are busy "digesting" the carbon.

In humid summer areas, a mulch of white sunlight-reflecting sand beneath and around lavenders helps ward off mold and fungus attacks, while at the same time increasing the volatile oil content of the plants. A gardener who once worked at the CIA brought home shredded documents and achieved an effect similar to white sand. In other areas, with other plants and conditions, a whole different approach to mulching may be appropriate.

However, when I mention "mulch" in this book, I'll usually be referring to spoiled hay, straw, newspaper, and cardboard. Hay is the common term for various plants (such as alfalfa) grown and cut specifically for animal feed. The first cutting of hay in a season can contain a lot of seeds and create a weed problem, so try to get the product of a second or even third cutting. Hay is a lot more fertile than straw, but the trade-off is that straw contains many fewer weed seeds.

Straw is the stem left behind after a grain—such as wheat, oats, rice, etc.—has been harvested. Wheat straw can contain some hitchhiking seeds and may have to be weeded occasionally or just layered with more mulch. Where I live, it's fairly easy to get reasonably priced bales of rice straw (rice is grown inland in the Central Valley north of Sacramento) that contains virtually no seeds. Rice straw is thus the preferred mulch for many gardeners around here.

Deep mulching is a very effective way to build fertility "from the top down," but many popular gardening books offered by East Coast publishers are inherently biased toward cold winters and warm summers with periodic rains and frequently high humidity. Under such conditions, straw or hay mulch can easily rot down each summer to add to soil fertility. It's a different story in the drier parts of the West, because of their summer heat and aridity. In such moisture-sparse climates, it can take two or more years for deep mulching to provide the same benefits. Overhead sprinklers can compensate, but in hot, dry climates these are counterproductive, since water loss from sprinklers can approach 50%, depending on the type of sprinkler and on the wind speed, amount of sunlight, percentage of humidity, and air temperatures.

Another reason for the difficulty of establishing deep mulching conditions in western states is the well-deserved popularity of drip irrigation. Drip irrigation is the best way to conserve water in any garden and is extremely popular in the west. Its ugly black tubing, which conveys water directly to individual plants or areas, is often hidden under a mulch for aesthetic reasons. Since the moisture is not widely distributed on the surface, you frequently wind up with large areas where no water reaches the top of the mulch layer to assist in decomposing it. When seasonal rains begin in the late fall, the temperature drops and decomposition of the mulch slows down, which is why deep mulching takes longer to produce its beneficial effects on soil fertility in the West. Good gardeners just learn to cultivate patience and work in the flow of seasons and at nature's pace.

## Newspaper

In places other than pathways, wood-chipped "patios" or ornamental trees, I mulch with another readily available resource: newspaper. If I apply it carefully every year around perennials, vegetables, and shrubs, I don't have to worry about weeding until fall, at which time I simply add another layer of newspaper and mulch.

### ❧ PRACTICAL TIPS FOR GARDENERS

The use of mulch can be a double-edged sword. It can be invaluable for conserving moisture and for rotting down to create a rich, loamy soil. But, as mentioned, mulches can harbor any number of pests—mice, slugs, snails, and earwigs, to name just a few. No gardening method is without its downside, but the benefits of mulch are worth adapting to different situations.

Some gardeners I know use the "scorched earth" approach—completely bare soil for four feet in all directions from their raised vegetable beds. Others mulch with carefree abandon and deal with pests as they arise. Well-mulched gardens usually work best when the propagation method used involves transplanting into the beds, rather than seeding them directly. Of course, there are a few exceptions: the leaves of garlic, potatoes, shallots, and onion bulbs will easily grow through a straw mulch.

Organic mulches such as hay, straw, and shredded leaves not only prevent moisture loss, but slowly break down into a very mild fertilizer. To have any worthwhile effect in saving moisture, a mulch should be at least two inches thick. (If your moisture-control layer is more than four inches thick, you're probably wasting your time and money, except in the case of exceptionally loose mulches, like straw and spoiled hay.) Thicker mulches, however, are usually necessary to suppress weeds. Keep the mulch six or more inches away from the stem of each shrub or tree to ensure that the upper part of the root system doesn't rot.

## Recipe for Newspaper Mulch

Fill a five-gallon bucket nearly full of water and grab several sections of newspaper.

Nowadays, most black-on-white newsprint is printed with soy-based inks and no longer carry lead. (But, take those glossy ads to the recycling center, as heaven knows what inks and chemicals are used in them.) Dunk the newspapers into the water and swish them around for a few seconds, just enough to moisten them and not enough to soak them to the point where they're falling apart. Moistening the paper prevents the wind from blowing it away before decorative mulch is applied as a second layer to conceal it. Don't bother weeding unless the weeds are particularly high. If they are, cut them off at ground level and leave the cuttings to decompose beneath the newspaper.

Apply a layer of the moist newspaper four to five sheets thick around the plants (I bet you can't do this without stopping to read at least one article you missed or forgot about), being sure to overlap them at least four to six inches so that there are no in-between spaces for weeds to come crawling out. Don't bring the newspaper right up to the base or trunk of plants, especially drought-tolerant species, as this might keep in too much moisture and lead to the dreaded root rot (*Phytophthora* spp.; see Appendix #6).

After the desired area is covered with newspapers and looking totally trashy, it's time to hide the paper with an attractive mulch. I use old rotted rice hulls or the pen sweepings (wood chips mixed with poop) from a nearby turkey farm. Both are weed-free and are not too expensive if you get a big load delivered; one 15-cubic-yard truckload, if the extra is kept under cover, may last you for several years. Clippings from the lawn, if not layered too deeply, are an excellent way to cover the newspaper. Compost works OK but usually contains seeds of its own, essentially defeating the purpose, so if you're adding compost, do so before laying down the newsprint.

The mulch should be applied just thickly enough to cover the newspaper. Once this is done, I don't have to weed again for the whole dry California summer. In places with summer rains (which can mean floating weed seeds), I suspect you'll have to weed (though less than usual) or reapply the newsprint and mulch layer at least once. Eventually, the paper breaks down and becomes another part of the humus. In my Northern California garden, where we get no summer rains and I don't irrigate my xeric (dry) plantings, pieces of dried newsprint tend to sift to the top of the mulch layer by late summer. I either pick up the pieces or add another layer of newspaper over the scraps of newsprint and apply more mulch.

## Cardboard

Cardboard is especially useful for killing off large areas of lawn or weeds before planting and/or using the newspaper technique. Cardboard boxes used for shipping refrigerators and other appliances are ideal for this as they cover so much area. As with the newspaper layers, be sure to overlap the edges by at least six inches to control wandering weeds. Use one sheet of heavy-duty corrugated cardboard per layer or several if you're working with thin cardboard. Wet it down to hold it in place before covering it with the decorative mulch. A friend of mine inherited a lawn composed of some grasses and "weeds" like dandelions. She covered the entire 15' by 30' area with cardboard for two years and completely eradicated all the growth. When she cultivated the area, new seeds were brought to the surface and sprouted. Since she loves to weed, she didn't follow up with newspapers, preferring to pull the tiny seedlings up by hand.

## Torch Seedling Weeds

There is another way to deal with young weeds—use a propane flame to kill young seedlings. With this method you're actually boiling the water out of the tissues; no need to fry them to a crisp—just a quick pass of the flame is all that's required. This works especially well with monocotyledon seedlings. (Monocots, such as grasses, grow with parallel veins running along their leaves.) It's also a very useful way to avoid using herbicides on unwanted seedlings that pop up in cracked concrete walkways, patios, and other hardscaped areas. In areas that receive summer rains, you may want to try using a special propane torch to burn back the young sprouts.

Organic garden suppliers list products such as the Red Dragon Weed Torch™, which features a one-inch opening for work in tight areas, like near the edges of flower beds. You can also get a larger model for a blast of flame that will whack back intruding weeds in larger areas. Be sure all torching is done after a rain or while the mulch is thoroughly soaked. You should have a hose nearby and perhaps a friend to help stand watch as you keep an eye on smoking embers. Be careful not to singe nearby plantings; again, just make a quick pass over the young weeds to boil the water out of the tissue—the mulch will usually not catch on fire. See Figure #4 on the next page for an "industrial-strength" weed controller, which uses a heavy propane tank. Some gardeners prefer to use a small dolly to cart the tank around.

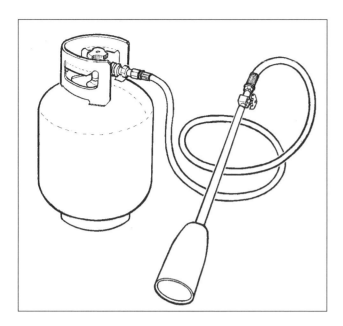

**Figure #4:** This larger propane-fired weed controller is dangerous; use with caution. It is very heavy and most gardeners use a small dolly to move it around. It's also *very* loud.

FOOTNOTE

Some soil experts estimate that a gram of soil (equal to the weight of a paperclip or the size of a cube of sugar) contains 2.5 *billion* (yes, billion with a "B") bacteria—as many as 20,000 species of them, as well as 400,000 fungi, 50,000 algae, and 30,000 protozoa. It's been claimed that the typical microbial population of a teaspoon of soil is greater in individual numbers than the human population of the earth.

FOOTNOTE

# CHAPTER 3

# Lawns

Lawn. Rhymes with yawn. Perhaps not the most thrilling way to start a chapter on the subject of roots or anything else (except possibly a civilized game of croquet), but the fact remains that, at least in this country, turf rules. By some estimates, 14–27 million acres of lawn blanket America. An area of cropped grass has no equal when it comes to outdoor activities like football, soccer, badminton, croquet, backyard campouts, laying down to watch for shooting stars, necking, and other time-honored pastimes both quiet and boisterous.

Lawns actually have many advantages over other forms of landscaping. One of the most important of these is that they actually take less time per square foot to care for than any other type of maintained planting. Consider the following list, based on hours of care needed per year for each 100 square feet of the plants in question:

- Roses = 17 hours.

- Annual bedding plants = 3 hours.

- Wisteria = 4 hours.

- Grasses (turf) = one hour or less.

[A study by the University of Tennessee Cooperative Extension Service of .8 of an acre public lawn showed that with a 42-inch wide mower, all mowing, soil aeration and applications of lawn chemicals (definitely not an organic lawn approach) took less than six or seven minutes per year (32 weeks due to winter) per 100 square feet.]

So, when you're on your knees cussin' up a storm because you can't get the lawn mower to start, just remember that a lawn still takes less time and effort to maintain than most other gardening choices.

## Know Your Roots...

Growth-wise, lawn statistics are mind-bogglingly impressive: It's estimated that the root system of a single one-year-old ryegrass plant includes 372 miles of roots and 6,213 miles of root hairs. These roots may increase by as much as three miles per day, and the root hairs can add another 50 miles every 24 hours. Your lawn is hard at work even when you're kicked back in a recliner with a mint julep.

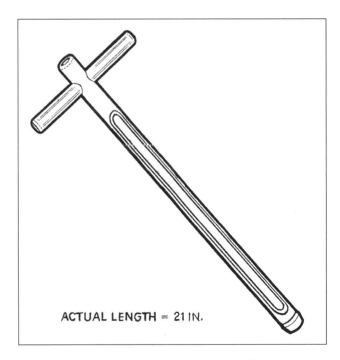

ACTUAL LENGTH = 21 IN.

**Figure #5:** This core sampler tool will help you fathom your soil without your having to dig a hole.

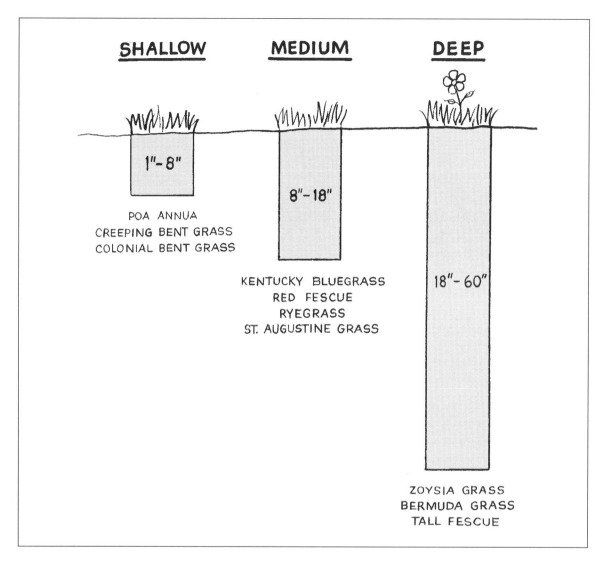

**SHALLOW**

1"- 8"

POA ANNUA
CREEPING BENT GRASS
COLONIAL BENT GRASS

**MEDIUM**

8"-18"

KENTUCKY BLUEGRASS
RED FESCUE
RYEGRASS
ST. AUGUSTINE GRASS

**DEEP**

18"- 60"

ZOYSIA GRASS
BERMUDA GRASS
TALL FESCUE

**Figure #6:** All turf roots are not equal. Yet, no matter how deep it grows, a majority of the water and nutrients absorbed by the root system comes from the top 6–12 inches of the soil.

## Know Your Soil . . .

A good lawn begins with healthy soil. Use a trowel to gather samples from several inches below the duff. Mix some of the soil from each location with water in a quart jar, shake the jar until all the soil is dissolved, and set it on a shelf for a week or more, until all the major components have settled out. (There will probably be some plant stuff floating on the top of the water; just ignore it.) At the end of that time, you'll find three distinct layers: the sand will be on the bottom, the silt (a mixture of particles of sandy loam and clay that are light enough to be easily carried by water, but not as slick as wet clay) will make up the next layer, topped by the clay layer. The relative thickness of each will

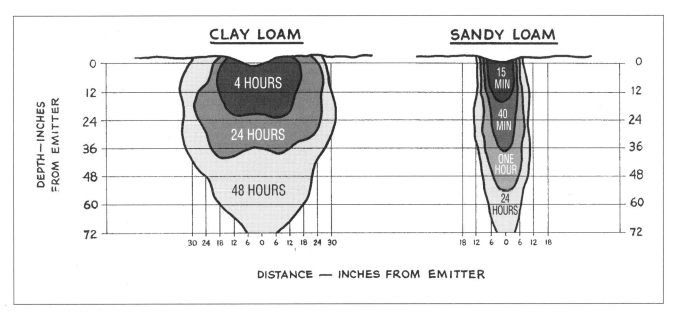

**Figure #7:** This drawing shows how widely water spreads in different soils over time. This pattern is much like that produced by a drip irrigation emitter on similar soil.

give you a pretty good idea as to the mineral composition of your soil. As mentioned earlier, the ideal mix is a loam composed of 45% sand, 40% silt, and 15% clay. Measure the layers in your jar and do the math.

Another approach to soil exploration is to dig two-foot-deep holes in several parts of the yard, fill each hole with water, then time how quickly or slowly the soil drains. If a hole takes less than an hour to empty, you've got pretty good drainage. If it takes longer than that, you might consider amending the soil structure with organic matter such as leaf mold and/or compost, or planting on a mound. [See the chapter "Planting Trees and Shrubs", page 129.]

Two variables that will determine factors such as fertilization and irrigation needs are: (1) the depth of the grass roots and (2) the soil type in your lawn. To literally dig up this information (don't be like a friend of mine who once said "You know, I love nature. I just don't want to get any on me."), make small trenches or holes in several places around your yard and measure for

yourself how deeply the grass roots penetrate. If you don't want to deface an established lawn, you can purchase a 20.5-inch soil-core sampler. [See Figure #5, page 23.] This is a narrow metal tube with a portion of the cylinder left open along its length to allow you to see the different layers of the soil in sequence from the top down. (They can also be used to tell how deeply a clay layer can be located.) Twist the tube clockwise into the ground. Twist counterclockwise just a bit and pull it out. With a little practice, you'll learn to spot the depth of the roots as they penetrate the core-sampler layers.

Of course, all turf is not equal. Bermuda grass, for instance, can grow roots up to eight feet deep in sandy soil, but its greatest mat of roots will be in the first six inches and the majority of them in the first foot down. (Be careful, however, about seeding Bermuda grass if you're not sure that's what you want; once it's growing in your soil, it's practically impossible to get rid of, as it will re-sprout from even the tiniest roots.) Bermuda grass is heat- and drought-tolerant and often used in the "Sun Belt." It is very rugged and quick to "repair" itself. St. Augustine, Bermuda, and zoysia are warm-season grasses with roots that grow quite deeply in summer and shallowly in the spring and fall. Kentucky bluegrass, perennial ryegrass, and tall fescue are cool-season grasses with deep root growth in spring and fall, shallow in summer. See Figure #6 for the relative rooting depths for various types of turf.

 **PRACTICAL TIPS FOR GARDENERS**

Before seeding or placing sod for a new lawn, be sure to amend the soil where appropriate and necessary. Once the lawn is growing, there's no easy way to access the soil for treatment or to improve the pore space.

If you're still cutting your lawn with an ancient power mower inherited from your dad or grandfather, consider investing in a new mulching mower. These up-to-date machines help convert grass clippings to humus by shredding the cuttings and blowing the "sliced-and-diced" leaves into the thatch (the lawn equivalent of duff), where decomposition can reclaim some of the nutrients and help develop good soil structure near the surface. Occasionally, if the soil in your yard starts to get too compacted from heavy foot traffic or other use, you may want to rent a lawn-plugging or aerating machine. You may also have to clear the thatch from the lawn periodically if it accumulates faster than decomposition can break it down. The necessity for this will depend on factors such as type of grass, frequency of mowing, amount of moisture, temperature, etc. You can choose to rent a de-thatching machine or take an old-fashioned rake in hand for some heavy-duty and productive aerobic exercise.

## Turf Root Irrigation—How deep is deep?

Many gardeners assume or have been told that the roots of lawn grasses grow three to four inches deep. While this is often the case, grass roots can grow as deeply as 17–18 inches, and, as mentioned, Bermuda grass can grow to eight feet deep in sandy soils. In most cases, however, you only need to irrigate the top 6–12 inches of the soil, as this is where most of the roots feed. [See Figure #6.]

One specialized way to get water to these roots is with subsurface drip irrigation (SDI). This is a system of drip-irrigation tubing that can be placed as much as ten inches below the soil surface. Because of the depth required, this is usually done during the creation of a new lawn or playing field, rather than as a retrofit to established stretches of turf. Small emitters are built inside the tubing at regular intervals; they may be 12, 18, 24, or 36 inches apart. The interval distance of the emitters along the line of the tubing and the distance between the rows of tubing as they are laid down are generally the same. These distances should be based on the type of soil you're working with. As an example, sandy soil requires the 12-inch spacing, while a heavy clay soil needs only 24-inch spacings. SDI is not the easiest watering system to install, as in most settings the tubing must be buried at least six to nine inches below the surface so that it is not punctured by plugging machines, which are used (especially in heavy-traffic venues like athletic fields) to perforate the ground for the purpose of aerating compacted soil.

 **PRACTICAL TIPS FOR GARDENERS**

SDI does have its advantages for special situations around the house. Buried in a strip between a sidewalk and the street, or along the base of a building, it's a

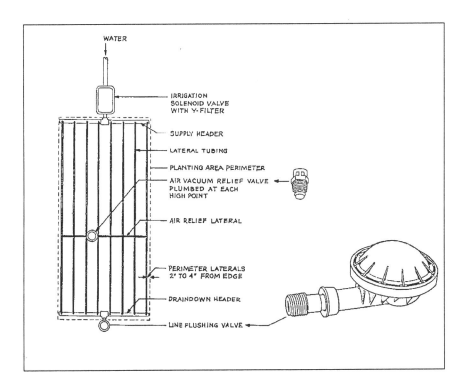

**Figure #8:** A brief overhead view of the main parts of a subsurface drip irrigation system (SDI). The water is evenly distributed to the soil and roots by placing the drip tubing at regular intervals. The only hose to use is in-line pressure-compensating emitter tubing that has the emitters built inside at even spacings. The emitters are designed to keep roots out. This one way to design a SDI. Another approach is described in Appendix #1, starting on page 139.

great way to irrigate flowers, shrubs, and trees without any unsightly tubing showing, while also reducing the chance of vandalism or accidental breakage. This system can also be used under a grass tennis court (if there are any left!), or in any other setting where you don't want to install fixed spray heads. (**IMPORTANT NOTE**: SDI will NOT work if your garden is infested with gophers or squirrels—they'll just eat through the tubing to get at the water.)

SDI is *the* most water-efficient way to irrigate lawns and turf, and is especially useful in dry climates (even though lawns in dry, desert-like climates are a horrible way to landscape and a major mistake— just my opinion).

See Appendix #1 for more about installing a SDI system. Figure #8 shows the layout of a typical SDI system.

Some of the water in an SDI system moves upwards by capillary action, and some moves laterally, but most will head downward, toward the main root systems of the grass. SDI has been found to be especially successful for irrigating stadium turf and has shown to be advantageous in some unusual situations For example, the frost line in Prince George, Canada, is approximately 10 feet deep, but Raymond Albers, of Advanced Irrigation Systems Inc., has maintained a successful 15-year history of playable turf by using subsurface tubing buried eight to ten inches deep in a football field.

## Watering

How frequently and deeply you irrigate will play a big role in your lawn's root health. Most references advise watering about one inch per week, depending on the weather. Others say a healthy lawn requires one to two inches of water weekly, including rainfall. It may be best to water once every two to three days during hot, dry weather, applying about 1/2 inch of water each time.

And right now you're probably asking yourself, How the heck do I know when I've applied an inch of water? Well, an inch of water-by-irrigation is equivalent to 62 gallons per 100 square feet. An easy way to measure is to "calibrate" your sprinklers. (You can do this and check their efficiency at the same time.) First, eat lots of tuna or moist cat food and save the cans. Collect a lot of them, remove the lids, and place the empty cans randomly around the lawn. (The bigger the lawn, the more tunafish salad sandwiches; the results will be more accurate with a large number of cans.) Turn on the sprinkler(s) until you can measure 1/4" of water in each can. Then multiply by 4 to get the length of time it takes your sprinkler(s) to distribute

one inch of water. (Or, just keep going and give the lawn a good watering until you've collected an inch of water per can.) If some cans get more water than others, you know you'll need to adjust the sprinkler head(s) for a more even distribution of moisture.

You can also simply poke around to see how well the water is soaking in. Use a trowel or a soil probe to check just how deeply the water is penetrating, and then adjust your water use, if necessary, for the grass type and soil conditions. As an example, sturdy Bermuda grass should be irrigated to the depth of one foot. Other turf grasses don't need such deep watering.

## ℞ PRACTICAL TIPS FOR GARDENERS

### Aerobic Intermittent Irrigation

As you've probably gathered by now, there's somewhat of an art to efficient watering. If you simply turn on your sprinkler system and walk away for an hour or two (or run it on an automatic timer), the upper portions of the soil may become flooded before the water sinks into the three to twelve inches your sod may require. Continuous flooding of the pore spaces in the soil produces an anaerobic condition that can kill the tiny root hairs needed for rapid absorption of moisture and nutrients. This is why the slow application of water by means of a drip-irrigation system is so productive—a much smaller area of the soil is anaerobic while you're watering, which means more of the soil "breathes" and the roots are happier. [See Figure #7 in this chapter.]

Older and more knowledgeable gardeners turn the sprinklers on for a short while, then shut them off for a few hours before applying a bit more water with a second watering. You can also use an irrigation timer that can be set to turn on and off at intervals. This approach to watering intermittently gives the lawn the moisture it needs while allowing the turf system to "breathe" between intervals of irrigation, which limits or eliminates runoff.

Another approach is simply to keep your eyes open and your senses tuned, and pay attention to your lawn. It's time to water the lawn when it begins to show signs of wilt. Walk across its surface; if your footprint remains depressed for several minutes (known to lawnmeisters as the "footprinting" effect), it's time to irrigate. Also, if your grass develops a blue-gray or blue-green color, it's very important to begin irrigation to avoid permanent wilting of the grass leaves. During periods of extreme heat and drought, dormant plants may be terminally stressed and die if irrigation is not altered to reflect actual need rather than watering by an artificial timetable. There's no substitute for getting to know your lawn's needs.

Watering is only a small portion of what a lawn requires. You can get plenty of good advice from a Master Gardener at your local Cooperative Extension. Cooperative Extension offices are usually listed under the County pages of your phone book. These state-subsidized organizations, assisted by Master Gardener volunteers, exist to address the needs of both home gardeners and commercial growers.

Being basically bureaucracies, Cooperative Extensions may know only the typical (read chemical) way to start and care for a lawn. If you want to banish all chemicals and maintain an organic lawn, get a copy of the classic on this topic: *The Chemical-*

*Free Lawn, The Newest Varieties and Techniques to Grow Lush, Hardy Grass*, by Warren Schultz (Emmaus, PA: Rodale Press, 1989).

# CHAPTER 4

# Weaver's Lone Prairie, or God's Very Own Lawn

Before we proceed any further, it's time to introduce Professor John Ernest Weaver (1884–1966), distinguished scholar and truly "hands-on" gentleman of the soil. Unbelievably patient and almost inhumanly persistent, Weaver spent decades searching out, mapping, and diagramming the root zones of plants ranging from native prairie forbs (broad-leaved herbs that grow along with prairie grasses) to shrubs and garden vegetables.

Weaver held the position of Professor of Plant Ecology at the University of Nebraska for 47 years. This close proximity to the great American prairie seems to have stimulated his already intense interest in the evolution of its complex ecology. Much of his meticulous research was focused on discovering just how the root zones of the intertwined matrix of prairie flora commingle. One short quote by Weaver says it all: "The prairie is an intricately constructed community. The climax vegetation is the outcome of thousands of years of sorting and modification of species and adaptations to soil and climate. Prairie is much more than land covered with grass. It is a slowly evolved, highly complex, organic entity, centuries old. Once destroyed, it can never be replaced by man."

Weaver worked closely with Frederic E. Clements (1877–1968), an ecologist with the Division of Plant Biology at the Carnegie Institution of Washington. Together, they coauthored *Plant Ecology*, Their book is the basis of this chapter on prairie flora. (Weaver also took a great interest in commercial vegetable crops, but that's in the vegetable chapter. No peeking!)

John Weaver literally went into the trenches to excavate the root zones of plants. Working and recording as carefully as the most compulsive archaeologist uncovering a buried civilization, he spent countless hours following and mapping roots and the patterns they made beneath his feet. The written and exquisitely drawn records of this tedious but horticulturally important work lay hidden in a dusty agricultural library at the University of California at Berkeley until the early 1980s, when I happened to stumble upon them. I sincerely wish that I could have met the man who spent so much of his time unraveling and diagramming the growth patterns of plants and their root systems. I often wonder if he found a wife who could understand and share his obsession with roots and the ecology of the prairie (or perhaps he courteously limited his working hours from nine to five).

It's curious that this master of minutiae rarely commented on how these root maps of the prairie and vegetable netherworlds might be put to practical use or relate to actual gardening practices. However, since the groundwork has been so marvelously laid, so to speak, I've taken it upon myself to extend the late professor's work by creating this book in "collaboration" with him, using as inspiration the drawings and commentary produced by his spirit, persistence, and labor.

One of Weaver's exhaustive studies involved buffalo grass (*Buchloe dactyloides*) as a native prairie grass habitat. This species originated as a native plant and is still found from Texas up through the great North American prairies.

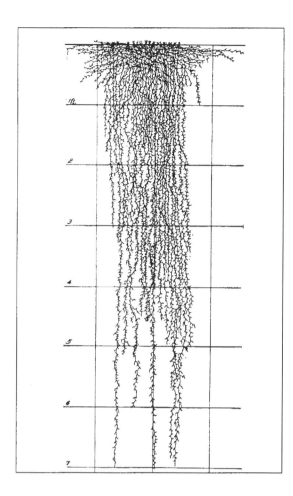

**Figure #10:** Buffalo grass can grow roots as deep as seven feet to utilize deep moisture. However, the most moisture and nutrients, if available, will be absorbed in the top 12 to 18 inches.
*From: Relation of Hardpan to Root Penetration in the Great Plains, J. E. Weaver & John W. Crist. Ecology (July 1922) Vol. 3, No3. Page 241. Grid equals one-square foot.*

Buffalo grass has been "domesticated" as a substitute for lawns in dry areas or for any place where water conservation is desirable. This hardy grass is well suited to the transition zones of the country, where it's often too hot for cool-season grasses (such as Kentucky bluegrass, perennial ryegrass, and tall fescue), and too cold for warm-season species (St. Augustine, Bermuda, and zoysia grasses).

As Figure #10 shows, Weaver excavated the roots of buffalo grass to a depth of seven feet. It must be the extensive depth of the roots that allows buffalo grass to withstand long periods of drought, although 70% of the root mass is located in the top six inches of the soil. Thus, as with many turf grasses, irrigation in those six inches is ideal for good-looking growth.

Figure #11 shows a cross section of mixed prairie grasslands. According to Weaver, "Bisects revealed the fact that (as with buffalo grass) the short grasses grow in silt loam so compact that surface runoff frequently causes the loss of over one-third of the precipitation....in sandy soils where the bluestem grows, practically all of the rainfall is absorbed, the soil is moist to depths of four to five feet, and to this depth it is penetrated by the deep roots of the tall grasses and other herbs," (*Plant Ecology*, 4).

## ✇ PRACTICAL TIPS FOR GARDENERS

Be aware that some lawn owners don't like buffalo grass at its most natural height of five inches because it's too long for walking or playing on. Lower cutting, however, may expose enough soil for weeds to invade. If your ideal lawn is short and putting-green velvety, you may want to opt for a type of grass adapted to that look. Ask your local plant nursery for suggestions.

To maintain its green turf, buffalo grass needs only .3 inches of water per week, as compared to .5 for Bermuda grass, .8 for tall fescue, 1.2 for Kentucky bluegrass, and 1.5 for perennial ryegrass. Another way to look at it is that buffalo grass can last 21–45 days without irrigation, as compared with St. Augustine grass, which needs watering every five days. Buffalo grass spreads by runner roots to slowly fill

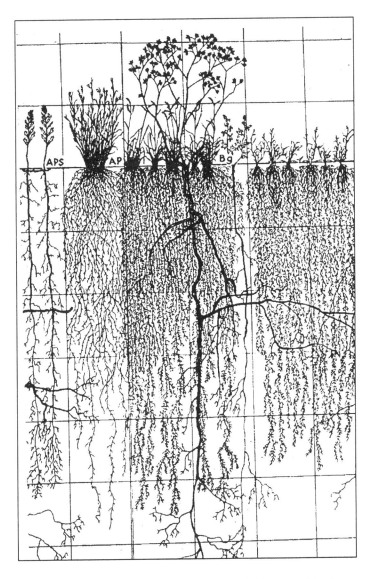

**Figure #9:** A bisect from west-central Kansas. Each square is one foot. "Bg" is buffalo grass. "Ap" is wire grass (*Aristida purpurea*). "APS" is western ragweed (*Ambrosia* spp.). The large shrub in the middle is a legume (*Psoralea tenuiflora*). *This drawing is used with the permission of the McGraw-Hill Company, Inc. From Plant Ecology, by John Weaver & Frederic Clements. 1938. Page 40. Grid equals one-square foot.*

Figure #9, a fascinating drawing executed in west-central Kansas, shows buffalo grass roots in their natural setting, mingled with other grasses and a lupine shrub. In its native habitat, buffalo grass often must deal with competitive plants. Weaver adds: "On the Great Plains, which receive less than 17 inches annual rainfall…[the] silt loam [is] so compact that surface runoff frequently causes the loss of over one-third of the precipitation. Usually the [top] 12 to 18 inches of soil alone are moist, and absorption is largely confined to this layer" (*Plant Ecology*, 41).

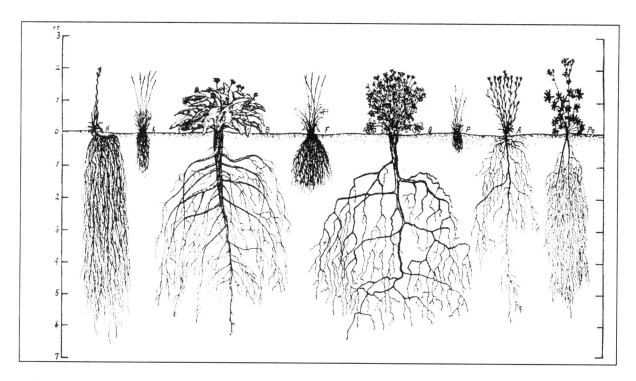

**Figure #11:** Prairie plants from eastern Washington. From left to right: hawkweed (*Hieracium* spp.), June grass (*Koleria* spp.), balsamroot (*Balsamorrhiza* spp.), blue bunch grass (*Festuca* spp.), Geranium, a bluegrass (*Poa secunda*), a composite (*Hoorebekia* spp.), cinquefoil (*Potentilla* spp.).

This drawing is used with the permission of the McGraw-Hill Company, Inc. From Plant Ecology, by John Weaver & Frederic Clements. 1938. Page 289. Grid equals one-square foot.

in a yard; in its native Texas prairie, with less than 20 inches of rain per year, it will form a nearly continuous cover. It seems to thrive in conditions where there are fewer plants for competition.

Prepare your soil well and as deeply as possible before planting plugs of (or seeding) buffalo grass. You won't have another chance to create conditions that will allow rain to penetrate and avoid losing precious moisture. Look for the more recent selections of buffalo grass strains (such as 'Legacy®' and 'Prestige™') that are available through suppliers (these tend to have been chosen for the easiest growth and most consistent reliability) and check with your local Cooperative Extension as to its appropriateness for your area.

To gauge your watering needs, you can use the original digital watering sensory device—your finger. Stick that digit into the soil in various places (or use a trowel to protect your manicure). In time, your touch and instincts will guide you as to how long to irrigate buffalo grass for proper moisture levels.

# CHAPTER 5

# Shrubs

Pity the poor shrub—caught somewhere in classification between lawns and trees, and its history and outlook on life so neglected.

The definition of a shrub is a bit vague, depending upon your source. Here is the online site Wikipedia's definition: "A shrub or bush is a woody plant, distinguished from a tree by its multiple stems and lower height, usually less than 20 feet tall. A large number of plants can be either shrubs or trees, depending on the growing conditions they experience." The Society of American Foresters defines a shrub as: "a woody, perennial plant differing from a perennial herb in its persistent and woody stem, and less definitely from a tree in its lower stature (size) and the general absence of a well-defined stem." You get the idea. Or not.

The problem is that some plants may grow as either trees or shrubs, depending on factors such as soil type, climate, and site conditions. Species we might usually think of as shrubs can sometimes grow to tree size. Examples of these include: boxwoods (*Buxus* spp.), hawthorns (*Crataegus* spp.), holly (*Ilex* spp.), bearberry, manzanitas (*Arctostaphylos* spp.), and rhododendrons and azaleas (*Rhododendron* spp.).

Some well-known shrubs include: barberry (*Berberis* spp.), daphne (*Daphne* spp.), forsythia (*Forsythia* sp), hydrangea (*Hydrangea* spp.), firethorn (*Pyracantha* spp.), currant (*Ribes* spp.), rose (*Rosa* spp.), and rosemary (*Rosmarinus* spp.).

## Growing Shrubs

Although little has been done to study the root systems of even the more popular shrubs, there is some fascinating shrub-related literature available for native plant root systems of the Southwest.

Ironically, the most detailed information on all root zones for grasses, forbs, native shrubs, and trees is found in studies done for the Los Alamos National Laboratory in New Mexico. The reason can be found in the titles of two research papers: "*Rooting Depths of Plants on Low-Level Waste Disposal Sites*" (1984) (and they're not talking soiled baby diapers!), by Teralene S. Foxx, et al; and "*Root Lengths of Plants on Los Alamos National Laboratory Lands*" (1987) by Teralene S. Foxx and Gail D. Tierney. Their concern in 1987 was, "…standards require operational procedures that will ban intrusion or disruption for at least 500 years without…the need for monitoring and maintenance of the disposal site after 100 years."

In the 1987 research, they were looking at "…a substantial earth cover…to bury…low-level nuclear waste". However, the study wanted to investigate the one variable that might put a stick in their spokes, that is, the depth of roots of native and introduced vegetation in the area. The 1984 study showed that "deep-rooted plants may provide a significant pathway for the release of buried toxic materials into the biosphere." The 1987 report also found that "…shrubs tend to show the longest roots in relation to overstory size."

The two Los Alamos studies do contain drawings of some roots, but I've condensed the rooting-depth visuals into Figure #12. In a xeric ecosystem (with plants adapted to an extremely dry habitat), plants survive by sending very deep roots, resembling taproots, to extraordinary depths. (I've included trees because

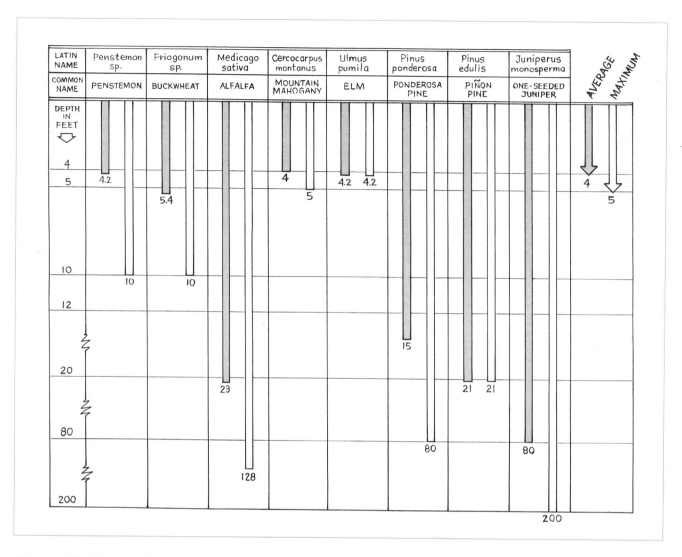

**Figure #12:** The rooting depths of shrubs and other plants as indicated in two studies at the Los Alamos National Laboratory.

they have the mostly likely root systems to penetrate a deep soil cap over radioactive waste.)

### ☭ PRACTICAL TIPS FOR GARDENERS

Don't put a toxic waste disposal site in your garden!

Use the same guidelines for buying containerized plants mentioned in Chapter #14, page 121, regarding native and ornamental trees.

Shrubs, if carefully chosen, can require as few hours as some lawn grasses in terms of hours per year for each 100 square feet per year. Examples include:

Shrubs requiring one hour or less per year:
- Camellia (*Camellia japonica*)

- Magnolia (*Magnolia* spp.) *Not* the tree *Magnolia grandifolia*.

- Oregon grape (*Mahonia aquifolium*)

1–2 hours:

- Boxwood (*Buxus* spp.)

- Juniper (Juniperus spp.)

- Azalea (*Rhododendron* spp.)

- *Syzgium paniculatum*

- Australian bush cherry (*Eugenia* spp.)

- Rockrose (*Cistus* spp.)

In the chapter on lawns, we learned that lawns need only six or seven minutes to one hour of care per 100 square feet per year, depending on the species of grass and the area. But, while it's easier and much cheaper to seed or roll out a lawn over large areas than it is to plant, mulch, and irrigate shrubs, especially in arid areas, it's a waste of a precious resource to use drinking water for lawn irrigation, as is done in many areas of the Southwest. (Reclaimed wastewater usually becomes available only with the development of new houses or subdivisions near a wastewater treatment plant.) Water use is becoming a large issue, politically and environmentally, so shrubs may soon necessarily replace lawns as sensible landscaping choices in dry areas.

A list of some popular shrubs can be found in Appendix #4.

FOOTNOTE

Rabbiteye blueberry plant (*Vaccinium* spp.) roots don't seem to vary that much from cultivar to cultivar. The roots of a 13-year-old bush can reach as deep as 31 inches. However, 96% of the dried roots were in the top 24 inches and 90% in the top 16 inches. In a heavier soil, 84% of the roots were found in the top four inches of the soil.

The root of the shrub live oak (*Quercus turbinella*) in Arizona is classified as having a "generalized" root system, one that features both a taproot and well-developed laterals. The taproot and other roots can penetrate up to 30 feet deep in fractured rock. At the same time, they have been found growing on soils only six to ten inches deep. Whatever the depth of the soil, the shrub live oak has a highly developed surface root system which gives it one leg up on grasses. The combination of shallow roots and deep roots give shrub live oak two sources of water–rapid flowing surface rains and deeper stored moisture.

Mountain mahogany (*Cercocarpus montanus*) has begun to show up in domesticated landscapes. The average rooting depth is 45 inches and the extremes run from 16–60 inches.

FOOTNOTE

# CHAPTER 6

# Vegetables

## ASPARAGUS

Being a perennial, asparagus has plenty of opportunity to develop a massive root system over time [See Figure #13], which helps it to live to be 50 years old or older. The example shown here is six years old, with roots that extend nearly eight feet wide and almost eleven feet deep.

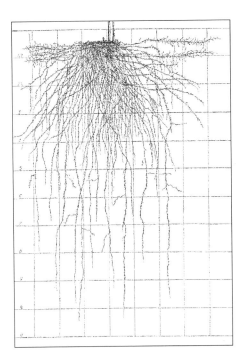

**Figure #13:** A six-year old asparagus plant, with roots that extend nearly-eight feet wide and almost eleven feet deep.
All the illustrations in this chapter are used with the permission of The McGraw-Hill Company, From *Root Development of Vegetable Crops*, by John Weaver & William Bruner. 1927. This illustration is from Page 62. Grid equals one-square-foot boxes.

[Each square is one-foot square, as are most of the rest of the illustrations in this chapter.] Although asparagus roots can potentially occupy over 500 cubic feet of underground space, their branching off is most abundant and active in the top foot of soil.

When grown from seed, the asparagus naturally develops a taproot a few inches long. This is often removed by commercial growers to cause the formation of thick, fleshy side roots, which are often found on the one- or two-year-old bare-root stock available in nurseries for transplanting. Many gardening books recommend planting one or two-year old asparagus bare-root stock in a trench up to one foot deep, with soil and manures added each year until the trench has been filled back up to the surface. (Don't harvest any asparagus for two years so the roots can build up some mass.) If situated in a good, deep, loamy soil, asparagus roots can extend down as much as three feet in the first year. As the plant grows over time, some of the older fleshy roots are replaced with new roots—a process called, by some, "lifting of the plants." This means that as the new roots arise above the older roots, it's as if the crown of the plant had been "lifted." Now, more gardeners are planting the crowns of the bare-root stock just deeply enough to be covered with soil.

The productive plant in Figure #13 was originally planted six inches below the soil's surface, but opinions as to ideal depth for new asparagus plants differ. In *Gardening Without Work; for the Aging, the Busy and the Indolent*, Ruth Stout references J. A. Eliot, who preferred to plant his asparagus right on top of the soil, emulating the plants that thrive naturally when birds have scattered their red berries. Keep in mind that this all takes place in the moist, humid summers of New England, and the probability of bird-spread asparagus is extremely low in arid areas.

Asparagus is often considered a "heavy feeder," and some growers recommend adding manure to asparagus beds each fall. This advice, primarily found in books published on the East Coast (and particularly in New England), is no doubt appropriate in places where the ground freezes. However, in areas such as Northern California, where it rains during the winter but very little in the summer, a fall application of manure can be a waste—it's likely to wash down below the roots or off of the soil's surface during very heavy rains. In such climates a spring application is best, ideally just before the rainy season ends, a matter of guesswork each year. Some gardeners maintain that asparagus is hard to overfeed, but Weaver has a different take on the subject: "When old roots die and decay, they furnish the soil a considerable amount of humus. The amount of humus is so great the continued use of manure as a source of humus is sometimes apparently not beneficial." (Weaver, *Root Development*, 69).

Autumn mulching with straw in areas where the ground freezes will help protect asparagus roots from too much cold by trapping the snow in an insulating layer. Leaving the tree-like tops of the plants to winterover similarly helps protect the crowns of the root systems. Extreme care should be taken when raking back the mulch as spring approaches, as it's easy to damage the tender tasty shoots as they begin to poke up from the soil. Once the edible shoots are finished producing and you've allowed some to grow up to form feathery foliage, bring back the mulch. In mild winter areas, mulch is not really required except to prevent erosion. When mulching and/or manuring an asparagus plant, be sure to remember that the diameter of the root zone can be six or more feet on a mature plant.

# CARROTS

When you pull a carrot out of the ground, you'll notice that it's covered with numerous tiny fibrous root hairs that need to be scrubbed or scraped off before eating or cooking. This might lead you to think of the carrot as having a very small, even shallow, root system. Nope. A carrot *is* a root crop, and as Figure #14 illustrates, a carrot plant can send its taproot as much as seven-and-a half-feet deep into a loamy soil.

In this illustration of a carrot growing naturally,

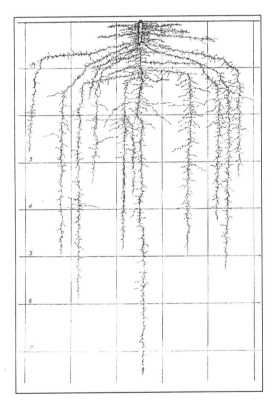

**Figure #14:** When you pulled that carrot from the soil, I'll bet you didn't know how many roots you left behind.
From *Root Development of Vegetable Crops*, by John Weaver & William Bruner. 1927. Page 210. Grid equals one-square foot boxes.

notice the dense horizontal roots in the top six or so inches of the soil. Weaver comments, "The greatest branching is in the surface two to four inches of soil, where a few laterals extend horizontally eight to ten inches…About the middle of August, maturing plants have well-formed "carrots" from which fine roots arise in great abundance. These furnish an excellent surface-absorbing system near the plant" (*Root Developmen*, 212.)

The growth of the carrot root in the upper two feet of the soil is somewhat similar to that of beets except that the carrot grows such a prominent taproot and produces numerous horizontal roots that can reach downward many feet.

## ℘ PRACTICAL TIPS FOR GARDENERS

It's obvious that carrots prefer a deep soil with good tilth and drainage, free of rocks and obstructions that can produce deformities. In less than ideal soils, cultivate as deeply as possible before seeding. Carrots are best grown in double-dug beds, or in boxes raised 24 inches above the soil line and constructed with wire bottoms to deter gophers or other underground gnawing pests.

When building garden boxes, line them with one-half-inch aviary wire, which comes in four-foot-wide rolls. One-inch chicken wire, while less expensive, may allow baby gophers to sneak inside the box. The aviary wire also has more galvanized metal and lasts longer in the ground. Even with this protective barrier in place, the taproot and many other roots will be eaten at the bottom edge of the wire.

The Cadillac version is to use one-quater-inch hardware cloth because it doesn't rust through as quickly.

When preparing beds for planting, be sure to work the soil with a flat-bottomed spade so you don't damage the wire.

When planting root vegetables, you can choose between dense spacing, which means maintaining lots of fertility, or planting further apart, and watering and fertilizing less frequently. As an example, consider the work of Steve Solomon in his book, *Water-Wise Vegetables*. Steve was living fifty miles south of Lorane, Oregon, and west of the Cascade Mountains when he did his research with two plots in his garden—one irrigated and the other "dry-farmed." His carrots (grown in a deep fertile soil) were thinned one foot apart in the row, with the between-row spacing at five feet. The carrots received no water whatsoever all summer. By season's end, the dry-farmed carrots had an average diameter of five inches and weighed over one pound each. In his experience, "The roots are not quite as tender as the Nantes types but are better than you'd think. Something about accumulating sunshine all summer makes the roots incredibly sweet" (Solomon, 58.) He has better results with the cultivars 'Royal Chantenay,' 'Fakkel Mix', and 'Topweight' than with the more common Nantes varieties. Solomon's experience is that spacing plants eight times further apart than the guidelines for intensive-bed cultivation produces about one-half the yields of plants spaced according to the guidelines.

# CAULIFLOWER & CABBAGE

Figure #15 shows how massive a cauliflower root system can be, even when the plant is only eight weeks old. Cauliflower customarily grows a taproot, but transplanting usually removes it, causing the plant to form multiple side shoots as shown. Here we see how important

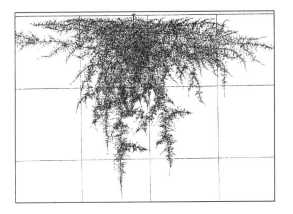

**Figure #15:** At only eight weeks old, the root system of this cauliflower plant is more than four feet wide and three feet deep.

From *Root Development of Vegetable Crops*, by John Weaver & William Bruner. 1927. Page 125. Grid equals one-square-foot boxes.

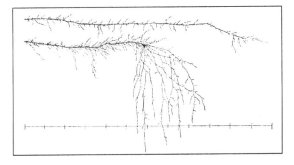

**Figure #16:** Shallow cultivation early in the season can produce more feeding roots as seen in the lower roots of cabbage in this illustration. The upper roots are from uncultivated soil. (Here the scale is in inches.)

From *Root Development of Vegetable Crops*, by John Weaver & William Bruner. 1927. Page 110.

the first six to twelve inches of soil are to the development of roots, and especially to those that tend to form tiny root hairs. Figure #16 (here the scale is in inches, and the plant shown is cabbage) shows how early shallow cultivation can actually promote branching of the surface roots. (The greater the number of "rootlets," the bigger the surface area available to absorb nutrients will be.)

Later in the summer (June 28 in this study) uncultivated soil beneath the leaves of the cabbage plant had developed roots to within *two millimeters* of the soil's surface! Many vegetable roots, and certainly those of cauliflower and other brassicas, follow this same pattern.

By July 19, the structure of the cauliflower root system has changed to that shown in Figure #17, with more roots extending into the second foot of the soil. Perhaps this expansion gives the roots more soil to draw upon for the nutrients needed to form the head of the cauliflower.

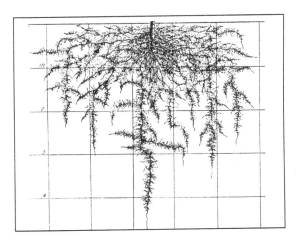

**Figure #17:** The scale returns to one-foot squares with this drawing of a mature cauliflower plant. Weaver says climate is more important than soil when it comes to the extent of the rooting. He also states that a constant supply of water is required.

From *Root Development of Vegetable Crops*, by John Weaver & William Bruner. 1927. Page 108. Grid equals one-square-foot boxes.

Early in the season, hand-weeding or very, very shallow cultivation, followed by plenty of mulch, is the best way to protect fine root hairs near the soil's surface while the plant is young. Weaver warns, "…In the case of cabbage (and cauliflower), late cultivation might do more harm than good" (Weaver, *Root Development* 127). This approach would've warmed the cockles of Ruth Stout's heart. Here is a plant that really thrives on deep mulching. Since the root system is a bit wider than the average width of the cauliflower's foliage, you should be sure to mulch beyond the area shaded by the edges of the leaves.

It's probably best to start with transplants that are big enough to survive the onslaught of root maggots, snail, slugs, earwigs, cutworms, and aphids. If your garden is full of some of these pests, it's best to leave the soil bare until the plants are well established.

If you have cutworms, a collar made from a waxed-paper cup helps. Cut off the bottom and make a radial cut outward from the hole. Turn the cup upside down and insert it into the soil as far as two inches. Or, use a tin can with both the top and bottom removed.

To discourage root maggots, try adding beneficial nematodes to the soil before planting.

# CORN

*As the [Cherokee] Indians sat about their fire, a cloud descended and from its midst a fair-haired maiden, dressed in flowing green, appeared to them. She was so lovely that one of the braves sprang forward to clasp her. She threw up her hands to repel him, and at the instant he touched her she disappeared. In her place stood a tall maize-stalk, its leaves her green gown, its silk her fair hair, its little roots her bare toes. A voice from the rustling leaves spoke to them:*
*"Clear the forest and plant the grain."*

From: *Roots, Their Place in Life and Legend,* by Vernon Quinn.

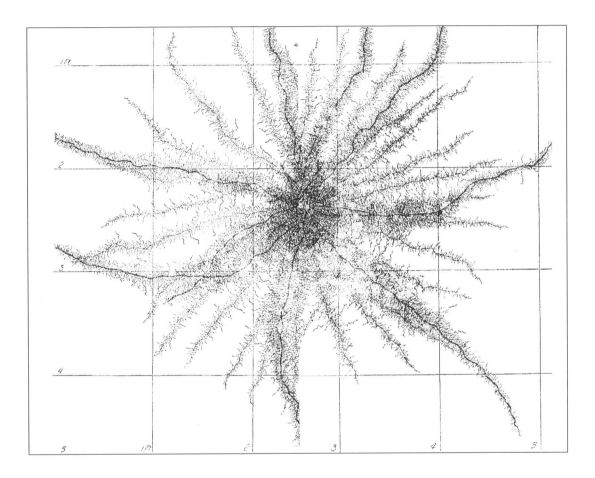

**Figure #18:** This beautiful diagram re-creates the pattern (seen from above) of corn roots growing in the top six inches of soil.
From *Root Development of Vegetable Crops,* by John Weaver & William Bruner. 1927. Pages 30-31. Grid equals one-square-foot boxes.

The roots of a mature sweet corn plant, when viewed from above, reveal a magical sight, a veritable living mandala. [See Figure #18.]

The exquisite detail of John Weaver's drawings (his working method is described in the first paragraph of the "Prairie Grasses" chapter) is revealed in this illustration, which depicts corn roots found in the top six inches of soil.

The corn plant illustrated here is one of a crop planted on hills 42 inches apart with a distribution of three to four seeds per hill. Weeds were controlled by shallow cultivation (no deeper than one-and-one-half inches, using a "mulching fork," a farm implement that works like a hoe).

It should be noted that all the vegetable illustrations here were done from plants grown in a fine sandy-loam soil that had been manured for a number of years and had already been used to grow vegetables. The soil was prepared by plowing eight inches deep, then disked and harrowed to create a firm seedbed. (A disk is a tractor-drawn arrangement of disk-shaped plowing pieces that is used to turn over the earth in a field. A harrow is a tined implement that is dragged over previously disked land to crush clods of earth and level the soil.) The vegetables were grown with summer rains alone—no irrigation. All the straight lines in the illustrations form boxes that represent one square foot.

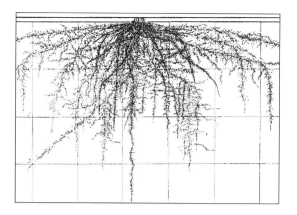

**Figure #19:** Corn can grow quickly in a good soil. This view is of an eight-week-old corn-root system.
From *Root Development of Vegetable Crops*, by John Weaver & William Bruner. 1927. Page 26. Grid equals one-square-foot boxes.

Corn plants produce a massive root system that consumes large amounts of moisture and nutrients. This root system forms quickly; by the time a corn plant has sprouted just eight leaves, it has produced 15 to 23 main roots with a total of 8,000 to 10,000 lateral roots. A mature plant can generate roots that have "ramified" (grown through) as much as 180 cubic feet of soil. Figure #19 shows a crosssection view of an eight-week-old corn-root system. Notice the substantial number of main and lateral roots at

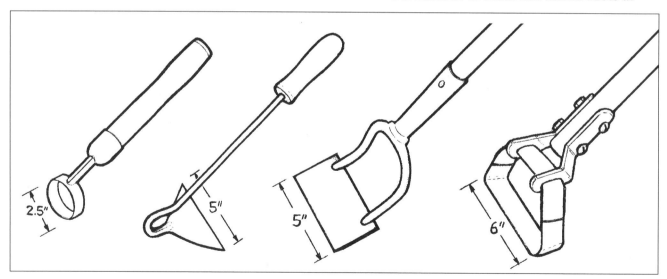

**Figure #20:** A selection of tools to use for surface cultivation. From left to right: Henningson Circle Hoe™, scrapper (Dutch hoe), a weed skimmer (hoe), and a Hula Hoe™. The Hula Hoe™ works when pushed and/or pulled.

the level of one foot or less. As a rule of thumb, Weaver quotes other research that says: "Briefly, sweet corn roots of corn extend laterally more than half as far as the stalk extends upward, and the root depth is equal to the height of the stalk..."

Weaver also cautions against cultivating more than one to one-and-one-half inches deep around corn plants for weed control because "Even shallow cultivation cuts many of the roots, and deep cultivation is very harmful and greatly decreases the yield. During a period of eight years, the average yield of corn in a cultivation experiment in Illinois was 39.2 bushels [when] cultivated 3 times; 45.9 bushels where no cultivation was given but the weeds were kept down by scraping with a hoe; and a mere 7.3 bushels…where the weeds were allowed to grow." Though I doubt that most gardeners plant corn by the acre, the relative impact is notable.

Weaver adds that corn roots absorb nitrates at all levels of the root system. He cautions, however, against adding manures to the "hill" (a practice where corn is planted on a small mound), as it promotes early growth but provides little benefit at the time of the formation of the corn ears. In his observation, those plants that were fertilized at the hill had smaller root systems and were more susceptible to drought damage and reduced yields. Weaver thus deduces that localized application of fertilizers also localizes roots and prevents them from naturally ramifying a larger volume of soil.

## ☙ PRACTICAL TIPS FOR GARDENERS

I see the corn-root system as a good model for using a different set of tools—implements more appropriate to gardening than to farming—for surface cultivation. There are three effective ways to skim the roots of weed seedlings without cultivating as deeply as with a regular hoe (See Figure #20.) One is the use of a so-called "Dutch Hoe," which can scrape as deeply or shallowly as the gardener wishes and works by pushing the hoe forward. Another option is the "Hula Hoe" (also called the Action Hoe™), with which it's possible to scrape down less than one-and-one-half inches, and which has the advantage of being able to cultivate in both directions. The Henningson's Circular Hoe™ makes it easy to keep cultivation near the surface while aerating the soil. All of these tools can be used without damaging the sensitive and vital portion—the upper four inches—of the corn's root layer. Deep mulching of corn will suppress weeds, keep the soil cooler in hot summer areas, and allow the roots to grow upward to the very surface of the ground, perhaps even up into the mulch.

### WARNING: A Personal Sermon

Corn needs lots of nitrogen, and leguminous plants, as long as they're not too crowded, can provide it with all the nitrogen required. In a "natural" garden, the ultimate goal would be to eliminate all imported nutrients. Horse manure, cow manure, sacks of bone meal, blood meal, green sand, bat guano, phosphates, etc., all add additional fertility and qualify as "natural," but come with various environmental costs attached, such as mining, transportation, energy use, and wasted bulk.

As an example of the energy invested in nitrogen for corn, consider blood meal. According to the Food and Agriculture Organization's regulations, blood is introduced into the (processing) tank as a coagulated mass, previously obtained by a steam-action process. Ideally, as much liquid as possible should be squeezed from the coagulum. Heating is initiated at 82°C (180°F) and progressively raised to 94°C (200°F) for about three hours, then elevated to 100°C (212°F) for 7 hours. (That's a LOT of energy.) Drying is complete when the final moisture level in the dried product is about 12 percent.

No matter what your source of imported nitrogen, whether for soil preparation or as a summer application for growth, it's most effective to spread it relatively far from the cornstalk itself in order to feed the massive width of the corn-root system most efficiently. One method of doing this would be to fertilize between the rows rather than on the rows themselves or at the base of each plant. If you plant intensively, be sure to add plenty of nutrients for this hungry crop.

Purplish strips at the edges of corn leaves indicate a deficiency in phosphorus. Some organic gardeners use colloidal phosphate as the solution to this deficiency, but consider this: colloidal phosphate is often strip-mined in Florida, washed with water (and Florida has a *big* problem with supplies of fresh water), loaded on train cars, and shipped to places as far away as California and Washington, where it's sacked up and shipped to your local garden-supply store. And the total amount of phosphorus ($P_2O_5$) in the sack is only 16 percent of all the bagged-up bulk, of which a mere two percent is available the first gardening season since the phosphorus is locked up in a mineralized form that requires the activity of soil microbes, soil bacteria, and exudates. [The action of root exudates was explained

earlier in Chapter 2.] Unnecessary water use, exploitation of limited resources and wasted energy—all wrapped in a single bag. Add to this the fact that there is a limited supply of easily mined colloidal phosphorus. It's much like oil: will we have enough in the future? Will we be able to find enough new supplies if the current mines are exhausted? Some say the U.S. supply will be gone by 2035.

Choosing to buy commercial colloidal phosphate and blood meal really means making a very important environmental decision. This is especially clear when one compares the environmental cost of imported amendments to the energy-efficiency of "growing" nitrogen and phosphorus at home by planting legumes and tilling the young foliage into the soil. Thus the gardener has two choices: (1) import nutrients, organic or not, to force an intensive yield, or (2) use wider spacing when planting and/or rotate crops to cut down on the competition for available nutrients. Whichever you choose, don't grow corn in the same spot every year, as it will exhaust much of the nitrogen. Instead, alternate corn crops with green manures—legumes tilled into the soil to provide nitrogen from the atmosphere and increase available phosphorus. See Figure #21 for more on when to turn under a green manure crop. [For more about green manures, see Appendix #2.]

Consider the cultivation of corn in the arid southwest. The Hopi Indians, from centuries of experience, know that corn roots are extensive and greedy. They plant hills of corn with multiple seeds, but the hills are widely spaced to prevent depletion of soil nutrients and conserve moisture. This is less of a problem in humid, moist, temperate areas, where nitrogen recycles faster, due to the decomposition of plant matter that is retained in the soil, and to greater rainfall. The trade-off of intensive imported fertilizers versus wider spacing still applies.

These corn-root "maps" show how extensive the roots can be and why the thoughtful gardener makes a careful choice between vertical (intensive) versus horizontal (wider spacing) cultivation. **End of Sermon.**

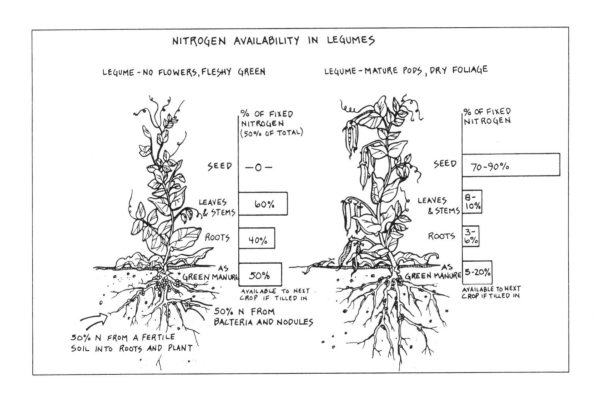

**Figure #21:** This illustration shows how important it is to turn under a fresh green manure crop before there are many blossoms. A legume stores most of its nitrogen in the roots and foliage before blooming—getting ready to move the nitrogen to the developing seed. The more tender foliage provides a greater volume of nitrogen and it more readily decomposes when compared to the more mature, "woody" plant with ripe seeds.

From *Designing And Maintaining Your Edible Landscape - Naturally.* by Robert Kourik, Metamorphic Press 1986. Reprinted in 2004 by Permanent Publication, UK.

# LETTUCE

As shown in Figure #22, lettuce has an extensive root system. This is why it's possible to harvest multiple cuttings of this classic and classy vegetable for salads, as long as a portion of its crown remains intact. The roots of lettuce in your own garden may not be as extensive as those shown in the illustration, which is of a two-month-old plant.

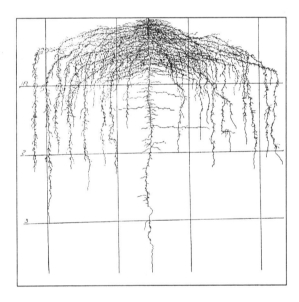

**Figure #22:** This is the plant used for the "test" on page three in the introduction—another example of how the small foliage of lettuce above ground misleads one about what grows beneath.
From *Root Development of Vegetable Crops,* by John Weaver & William Bruner. 1927. Page 324. Grid equals one-square-foot boxes.

## Lettuce Proceed...

Lettuces of all varieties have long been grown and used together as ingredients of cut-and-come-again salads. The older, more proper name for this reckless mingling of lettuce and other greens (which originated in France and Italy) is "mesclun," meaning "mixture." In the words of Rosalind Creasy (author of *Cooking from the Garden)* the original forms of mesclun were, "…intended to use every part of the tongue. European mesclun was, and is, meant to be a combination of seasonal greens with textures ranging from crispy to velvety and flavors ranging from tangy to bitter." Rosalind is not enamored of the typical American salad. "Americans do not eat a true salad. Our salads are the 'white bread' of greens, based on pale iceberg lettuce, and very limiting visually, flavor-wise, and especially nutritionally."

## Mesclun—drug or food?

The greatest problem initially faced by U.S. growers and markets in marketing "mesclun" salad greens was a confusion of its name with that of the illicit drug "mescaline." [The two are pronounced almost identically to the uneducated ear.] As a result, the equivalent of mesclun we often see in the supermarket is now usually called a "spring mix," which happens to show up nearly every day of the year…go figure.

A salad mix can contain numerous combinations, with the blend of greens constantly changing, often to reflect the changes of the season. Mark Musick, who worked at Pragtree Farms (a notable producer of salad mix) near Arlington, WA, estimates that, "By 1985, we used over a hundred varieties of greens and 25 or more types of edible flowers [in our salad mixes] in the space of a year."

## PRACTICAL TIPS FOR GARDENERS

A quick, simple, and efficient mesclun crop-for-harvest requires broadcast (scattered by handfuls as opposed to individually placed) seeding and frequent cutting. Some seed catalogs offer ready-to-plant mesclun/spring-salad seed blends; however, if you plant all the salad-green varieties, including lettuces, together, you'll only be able to harvest everything a few times, at best. This is because each type of seedling grows at a different rate, and each type reaches its optimum harvesting stage at a slightly different time.

The alternative is planting each ingredient for the mix separately. Choose the varieties of greens you want to grow, then buy and plant separate amounts of seed for each. Planting small areas with only one type of green means that you can cut and re-cut many more times than with an area planted to a mixture of seeds, because you can suit the harvest time to each green's requirements.

Other than a variety of colorful and tasty lettuces, some of the greens that can be harvested up to six or eight times include: arugula (*Eruca vesicaria*); 'Mizuna' (*Brassica rapa* var. *nipposinica*) or Japanese mustard (*Brassica japonica*); Russian red kale (*Brassica oleracea*, Acephala Group); white mustard *(Brassica hirta)*; and garden cress *(Lepidium sativum)*.

Doug Gosling, who for over 25 years has tended the garden at what is now the Occidental Arts and Ecology Center (OΛEC) in Northern California, says, "Our approach to soil cultivation is simple: before each bed is planted, we sprinkle it lightly with quarry dust, which costs only $9 per ton and provides valuable trace minerals. For each 100 square feet of bed, we apply two wheelbarrow-loads of compost and one wheelbarrow-load of aged dairy manure. Next, we single-dig the bed [Ed. note: The soil at the OAEC has been double-dug for over 15 years with soil amendments such as compost and manures, and, in some places, no longer needs double-digging], and then tilth (break up) the surface clods as we shape the beds." The final step before seeding is, in Gosling's words, "to 'massaging' the beds, using our hands to crumble all the small clods." With a droll, yet heartfelt look in his eyes, he's quick to add, "This certainly keeps us in touch with the earth!"

The easiest way to plant the seed is to lightly broadcast it, with the goal in mind of winding up with plants on two-inch centers. Next, water the beds immediately and then once a day, depending upon the weather, with the idea of keeping the soil moist, but not soaked, until the seeds germinate. Thereafter, again depending on the weather, you should only need to water them once a week. The beds usually need only one weeding while the seedlings arc young and one thinning to achieve the desired spacing. Naturally, any thinnings can go into that week's salad mix.

When cutting for use, carefully work your way through each patch of foliage, using a pair of lightweight kitchen scissors to cut each leaf. Be sure to leave one or two inches of foliage, as well as the apical (top center) bud, and several small, immature leaves surrounding it. By sparing these, you'll insure a repeat harvest in just a week's time.

Gosling explains it, "We're often growing the wilder end of the salad spectrum, which makes the plants hardier and innately resistant to bugs." No garden passes through all its seasons, however, without the trespass of an occasional pesty insect. The very process of harvesting allows the gardener to be selective and cut only those leaves that aren't infested. Or, as Doug Gosling says, "We just switch to harvesting different parts of the plants. If the leaves of mustard are attacked by aphids, we wait and harvest the tasty spicy young flower buds."

In the OAEC garden, located in the coastal environment 60 miles north of San Francisco, the insect pests of significance are flea beetles, root maggot, and, occasionally the diabrotica (or spotted cucumber) beetle, also known in garden slang as the "diabolical" beetle. Gosling's first line of defense is "to plant the most resistant varieties, which for flea beetles are: Russian red kale (*Brassica oleracea*, Acephala Group); orach *(Atriplex hortensis)*; lamb's quarters *(Chenopodium album)*; purslane *(Portulaca oleracea)*; quinoa *(Chenopodium quinoa)*; grain amaranth *(Amaranthus hypochondriacus)*; and sheep's sorrel *(Rumex acetosella)*." Since the flea and diabrotica beetles do the most damage to summer-grown brassicas, these crops become part of the mix only during the cooler times of the year and throughout the winter.

# ONIONS

The Southern white globe onion (*Allium cepa*), as shown in Figure #23 displays roots that grow less extensively than those of corn and further down from the soil's surface. John Weaver sums it up quite succinctly: "The relatively meager root system [of onions] and corresponding limited extent of above-ground parts explains why they can endure crowding better than most vegetable crops." In the case of onions, cultivation deeper than one-and-one-half inches once a week produced greater yields than the technique of scraping surface weeds—the reverse of the most efficient type of cultivation for corn. According to Weaver, the space of 14 inches between rows "…is quite ample for full root development with little or no competition between adjacent rows. In fact, the bulb crops are about the only vegetable crops for which this is true."

Some of the other root crops which display growth patterns similar to that of the onion are garlic, turnips, and beets (which, Weaver found, often maintained a root radius of 2.5 feet and a working depth of up to five feet, so that the first four to six inches of soil were not as important in terms of receiving water and nutrients). Radishes don't fit into this category, despite some of their roots growing as deep as two to four feet, as "Most of the absorbing area lies in the surface two to eight inches of soil."

## ❧ PRACTICAL TIPS FOR GARDENERS

The Hula Hoe™ and Dutch Hoe can be used with these root crops a little more deeply than with corn. Another good tool is a long-handed Henningson Circle Hoe™. [See Figure #20.] The Circle Hoe not only cuts off weed seedlings but is designed to aerate the soil at the same time. The arch of the blade is shaped to cut away from

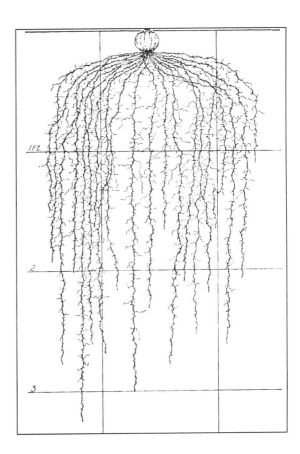

**Figure #23:** This root "map" of a Southern white globe onion is similar to that of other root crops such as garlic, turnips, and beets. From *Root Development of Vegetable Crops*, by John Weaver & William Bruner. 1927. Page 42. Grid equals one-square-foot boxes.

feeder roots and can be used in a side-to-side sweeping motion to cultivate between rows. All three hoes can be used for the shallow elimination of weeds—or to form a "dust mulch" when needed.

Since phosphorus helps increase the growth of roots and is essential to seed formation, it follows that root crops need adequate supplies of it. Here again, as mentioned in the section on corn, gardeners must make a decision between importing phosphorus or growing their own. Once established, nitrogen-producing legume plants [see Appendix #2] also provide available phosphorus, and the young legumes can also be tilled into the soil for the release of nitrogen, phosphorus, and other nutrients—a process called green manuring. According to E. W. Russell (Professor of Soil Science at the University at Reading, UK; in his book *Soil Conditions and Plant Growth, 10th Edition,* pages 278-279): "Green manures [grown] during wet off-seasons…often utilize less available forms of phosphate…hence an increase in the availability of [phosphate] for the crop."

Legumes usually make available enough phosphorus to sustain healthy crops. (You don't need the extra phosphorus if legumes already grow well in your garden's soil.) The various legumes can be tilled in once or twice, and then you may be able to return to surface cultivation if you desire. If corn plants show the purple discoloration denoting phosphorus deficiency, that will affect your onion crop as well, and you'll need to use green manuring again until the symptoms are gone.

## PEAS & BEANS

Garden pea (*Pisum sativum*) roots are pictured here in Figure #24. In the drawing, you'll notice a very prominent taproot growing to about two-and-one-half feet, while the lateral growth of the main root goes down further than the taproot and explores a diameter of nearly four feet. Most of the branching occurs in the top six inches of the soil.

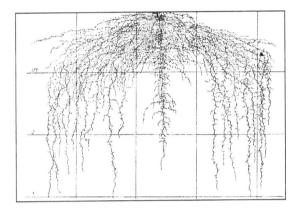

**Figure #24:** The common garden pea makes nitrogen-fixing nodules on this very extensive root system.
From *Root Development of Vegetable Crops*, by John Weaver & William Bruner. 1927. Page 177. Grid equals one-square-foot boxes.

The pea is a legume, which, like all legumes, forms nodules, or lumpy growths, along its roots. This is due to a magnificent symbiotic relationship between rhizobia bacteria and legumes roots. The roots provide some nutrition to the rhizobia bacteria. At the same time, the bacteria convert nitrogen gas into a solid form of nitrogen in the nodules. Thus it is said that these nodules "fix nitrogen from the air" and hold on to it for the plant's use. When the plant is grazed, mowed, or dies, the nitrogen becomes available to other plants. The grazing, mowing, and death of the plant (perhaps due to drought) "shocks" the plant, and the stress causes some of the nodules to shed from the roots and release nitrogen as they decompose. This can release as much as 40 to 250+ pounds of nitrogen, depending upon the variety of legume. The grazing (if not too severe) and mowing allow the plants to grow and be cut again to continue the fertility cycle. Weaver observed that with garden crops such as peas, the nodules are most often found in the top eight to sixteen inches of the soil. They can also be found "…irregularly distributed over the root system at depths of several feet."

To produce large nodules, field peas and other legumes need to be grown within a range of specific soil temperatures. The roots prefer an environment of $54^0$ to $75^0$F. In temperatures above $80^0$F and up to $86^0$F, the formation of nodules actually decreases. A soil that is too dry will also reduce nodule formation. If the soil already has plenty of nitrogen, the nodules may not form at all since there is already enough present to nourish the roots.

The kidney bean *(Phaseolus vulgaris)* is remarkably similar to the pea. Figure #25 is an aerial view of the top 6 inches of soil showing a kidney bean's root system, which resembles the fascinating mandala formed by the roots of a corn plant. The kidney bean rapidly forms a taproot when young. Before a young plant begins to mature, it produces a profuse number of roots in the top 10 inches of the soil, and the taproot grows to a depth of up to 24 inches. By the time the pods are forming beans, the taproot has increased its depth to three feet, with a working level of about 20 inches. The roots ramify the soil in a two-foot radius and three feet deep. While similar to a pea's root system, the kidney bean's roots grow more deeply and are more extensive beneath the plant.

Figure #26 shows how a mere lima bean plant can ramify 200 square feet of soil, with a majority of the roots growing and feeding in the top two feet of the soil.

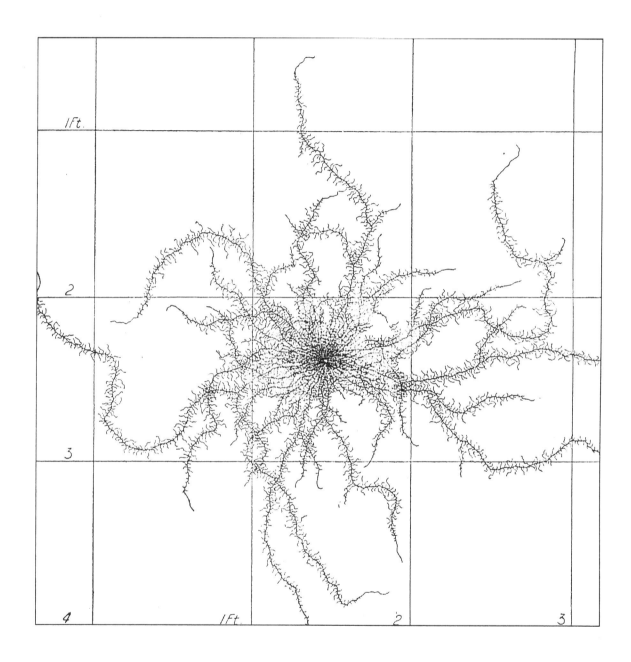

**Figure #25:** Another fantastic aerial view by John Weaver. This shows the top six inches of the root system of a kidney bean. (As seen on the cover of this book.)

From *Root Development of Vegetable Crops*, by John Weaver & William Bruner. 1927. Page 186. Grid equals one-square-foot boxes.

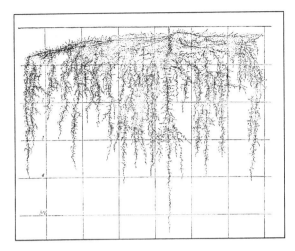

**Figure #26** shows the nearly mature root system of a lima bean.
From *Root Development of Vegetable Crops*, by John Weaver & William Bruner. 1927. Page 194. Grid equals one-square-foot boxes.

## ✃ PRACTICAL TIPS FOR GARDENERS

Mulch, of course. In cool spring weather, however, wait to do so until after the soil has begun to warm up, and make sure to mulch when the upper soil temperature approaches 80°F. Mulch at least one foot away from the base of the plant in order to cover the entire root system. In a "no-till" garden with a continuous cover of mulch, all will be well with the roots and their tiny root hairs.

Beans and peas are good crops with which to practice using newspaper as a weed barrier, followed with a better-looking mulch of straw to hold down and hide the newspaper. [See Chapter #3, page 20, "Humus & Mulch" for general instructions on mulching with newspapers.] After amending your soil or digging a young cover crop under, smooth the soil. If you're digging a cover crop under, wait two to four weeks before planting and mulching for the soil to decompose some of the raw plant matter. The soil's decomposers thrive on the new supply of nitrogen and will tie up the available nitrogen for a while until things settle down.

When you're ready to mulch, quickly dip a few sections of newspaper into a bucket of water; if you remember, this keeps the wind from blowing the paper around before the mulch is added. Lay down about four to six sheets of the newspaper with a four- to six-inch overlap to keep pesty vining weeds from sneaking around your newspaper. [This won't always work with weeds like sheep sorrel (*Rhumex acetosella*).] Poke holes through the newspaper at the interval recommended on the seed packet. Use your finger or a "dibble"—a tool used in Europe to plant seeds, transplants and bulbs. [See Figure #27, next page.] Insert the seed and cover with soil. Add a thin layer of straw mulch (or any other seed-free mulch you like) to conserve moisture. Cocoa bean hulls are the Porsche/Lamborghini of mulches in cost, but heavenly when sprinkled from above, filling the garden air with the aroma of chocolate. This is *not* a good mulch to use with drip irrigation; since the drip system is under the mulch, the surface stays dry and you'll miss out on the water-activated hedonism of the chocolate aroma. If you have a dog, be sure they don't eat any of the cocoa bean hull mulch, as chocolate tends to make dogs sick.

To help avoid greedy critters like snails, slugs, and earwigs, it may be best to start pea or bean seedlings in a pest-free area or buy at a local nursery and then transplant them. You may have to pull the mulch away from newly transplanted seedlings until they're hardy enough to deal with the ravages of insect terrorism.

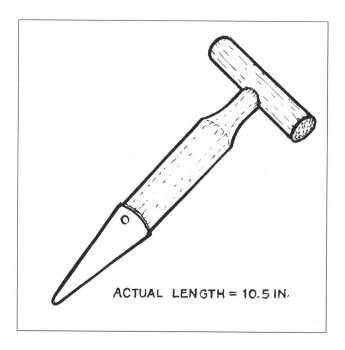

ACTUAL LENGTH = 10.5 IN.

**Figure #27:** Europeans have been using "dibbles" for hundreds of years to plant seeds, seedlings, and bulbs.

# PEPPERS

Figure #28 shows bell pepper roots as they looked when the plant was six weeks old. [Note: the scale on this drawing is six inches per side.] Figure #29—with a scale of one-foot squares—is a pepper plant shown when nearly full-grown. The striking observation here is how many of the younger plant's fine roots are growing in the top foot of the soil, near the surface. This plant's root hairs are quickly absorbing a vast amount of nutrients in preparation for the cycle of blooming and fruiting. Pepper plants are usually started in a sheltered place and then transplanted. Their normal tendency is to grow a taproot, but much of this large single root is usually destroyed upon transplanting. The plant responds by growing, from the remainder of the taproot, very numerous, profusely branched lateral roots. As the plant matures, its roots utilize deeper soils.

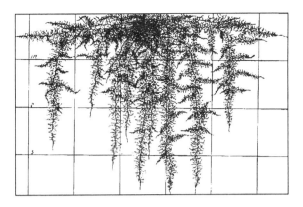

**Figure #29:** Here is the same pepper plant at near maturity. The scale is back to one square foot per square. The roots are now as deep as four feet, compared to two feet when the plant was only six weeks old.
From *Root Development of Vegetable Crops*, by John Weaver & William Bruner. 1927. Page 271. Grid equals one-square-foot boxes.

## ❧ PRACTICAL TIPS FOR GARDENERS

As with tomatoes, use seedling trays that air-prune the taproot to form a more fibrous root system. [See page 63.] At the early stages of growth, a *very* shallow surface cultivation will encourage even more roots and root hairs as the pruning of the roots causes more branching, resulting in more root hairs in the upper levels of the soil.

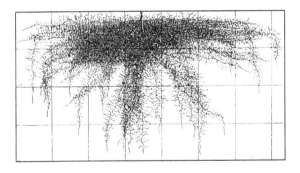

**Figure #28:** The scale here is six-inch squares. At only six weeks old, an amazing number of feeding roots of this bell pepper seedling are growing in the top six inches of the soil.
From *Root Development of Vegetable Crops*, by John Weaver & William Bruner. 1927. Page 270. Grid equals one-square-foot boxes.

# RHUBARB

Oh, did I love my great-aunt's rhubarb pie! Slightly tart, not too much sugar, and cut into generous slices you could really sink your teeth into. When rhubarb and strawberry pie appeared on the table, it was official: spring was in full swing.

The old rhubarb plants in my aunt's garden were enormous, and so was their appetite for manure. Looking at Figure #30, one can see why; as this drawing demonstrates, rhubarb can generate a massive root system, spreading as much as eight feet wide, with a depth of eight feet or more. It's easy to see why this plant needs a deep loamy soil to contain and support its large and hungry mass of roots and rootlets.

With rhubarb, as with most plants, the most active feeding areas are the root hairs at the ends of new growth. According to Weaver, "The thicker portions of the main laterals were nearly always poorly branched with only scattered young rootlets densely covered by root hairs" (Weaver, *Root Development* 71). This means that the older roots at the base of the stem were less important in absorbing nutrients than the newer shoots in the other areas of the root system. Providing nutrients in the area beyond the foliage [unlike the huge specimen in Figure #30, most rhubarb plants seldom reach eight feet across], helps encourage new roots and helps these roots explore and feed in new soil. The more wide-ranging the roots, the more fertilizer the plant will need. To investigate your rhubarb's root system, you can probe around with a trowel, using Weaver's map as a starting point.

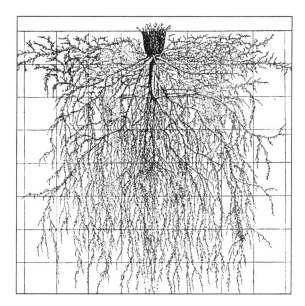

**Figure #30:** It takes a lot of roots to make rhubarb for one of my favorite pies!
From *Root Development of Vegetable Crops*, by John Weaver & William Bruner. 1927. Page 71. Grid equals one-square-foot boxes.

## ❧ PRACTICAL TIPS FOR GARDENERS

When fertilizing rhubarb, apply your manure or other nitrogen-rich fertilizer away from the base of the stalks and beyond the perimeter of the foliage, widening the diameter of coverage as the plant grows older. The most important area to fertilize is the zone of new root hairs near the outer edge of the foliage and beyond. The fertilizing and mulching should continue to expand with growth into an area perhaps half again as wide as the diameter of the foliage. In cold winter areas where the ground freezes, a deep mulch over the entire root system, especially over the "crown" of the plant, is a good idea. This will help to protect the plant from freezing too deeply and provide warmer soil in the early spring for earlier root growth (especially if the mulch is pulled away to take advantage of warm spring days).

When watering, follow the same guidelines as for mulching; to encourage the roots to ramify more cubic footage of soil, irrigate away from the crown of the plant and concentrate more on the perimeter.

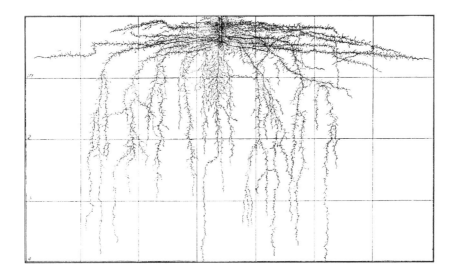

**Figure #32:** This is a drawing of a tomato transplant after only five weeks. The mature plant can grow roots up to 5.5 feet wide and nearly five feet deep.
From *Root Development of Vegetable Crops*, by John Weaver & William Bruner. 1927. Page 246. Grid equals one-square-foot boxes.

Ah, yes, tomatoes—one of the hoped-for star attractions of every summer garden. In Weaver's experience, a tomato seed planted in ideal outdoor soil, with no transplanting involved, can grow a taproot to the depth of 22 inches, at a rate of one inch per day. The tomato is yet another vegetable that prefers to grow a taproot, which is often damaged during transplanting. [See Figure #32.] This is why young tomato seedlings transplanted several times into increasingly large pots before their final move into the garden will probably end up with a root system more fibrous than that of tomatoes planted by seed in the garden. However, transplanting a tomato-plant stem deep into the soil will produce many adventitious roots along the length of the stem below the first leaves, creating a great root system early in the life of the plant. This is specific to tomatoes. Not many other transplants should be planted deeper than they are in the growing pot.

## ℘ PRACTICAL TIPS FOR GARDENERS

Tomatoes respond to transplanting from seedling trays to a larger, perhaps four-inch, pot, and yet again to a one-gallon "can." The best seedling trays are the ones that come in the form of connected rows of upside-down planting cones, each with a hole in the bottom—such as a Speedling Tray™ [See Figure #33.] As the seedling taproot hits the small hole in the bottom of the tray, it is "air-pruned," causing more side shoots and a larger number of root hairs. This prepares a vigorous fibrous root system and a long stem for planting below the soil surface to form those adventitious roots.

There are all kinds of ways to start plants well ahead of the time when it's safe to transplant them into the garden. These include cloches (bell-like glass coverings), plastic-tubing walls filled with water to catch and hold the sun's heat, and sunny windowsills or greenhouses. It's my observation, however, that large plants started well ahead of transplant time may not produce tomatoes any sooner than small seedlings planted in warm soil well after any threat of frost. (You can rake back mulches to allow the soil to warm more quickly.)

Onward to Surface Cultivation!

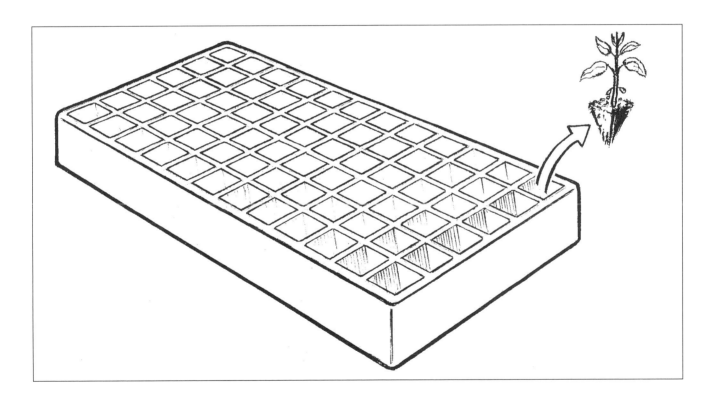

**Figure #33:** A Speedling Tray™, or one of the other brands of molded seedling trays with holes in the bottom of each funnel-shaped cavity, helps air-prune the taproot. This leads to more lateral, fibrous roots for healthier transplanting. The long pyramidal shape of a Speedling Tray compartment makes it easy to pull each plug of roots from the tray.

# CHAPTER 7

# Surface Cultivation & No-Till Gardening

By now, you've probably caught on to my not-so-hidden agenda of encouraging vegetable gardeners to experiment with surface cultivation or no-till gardening. No-till means never even disturbing the soil, while surface cultivation involves slicing the roots of weeds off just under the soil's surface—say two to four inches. As you've read in previous chapters, these methods certainly control weeds while encouraging root growth, healthy plants, and good yields.

Remember, however, that gardening guru Ruth Stout stoutly tilled her garden plot for 14 years before switching to her till-free technique, which involved applying a deep mulch of spoiled hay. So, go ahead and till your soil—even double-dig it—if the ground is clayey, has a hardpan barrier near the surface, or in any situation where a few more years or seasons of dig-and-till will contribute to a soil condition that requires less frequent tillage while maintaining good yields.

There are plenty of advocates for no-till gardening. Many say it mimics nature's way of improving soil from the top down, while others maintain that they're relying on earthworms to do the "tillage" at the speed of nature. An interesting fact is that artificial tillage can actually impede the storage of moisture in the soil. Research by Dennis Linden, soil scientist at the USDA Agricultural Research Station in St. Paul, MN, found that the common earthworm (*Aporrectodea tuberculata*) working under tilled soil forms a "less desirable pattern (of) meandering tunnels"

which slows the absorption of moisture by the soil. Worms under plots with no tillage, however [in this study, managed with herbicides], formed "a network of vertical burrows, due to night crawlers, and depressions, due to common earthworms, [which] tends to funnel water rapidly downward through [and into] soil layers."

According to Linden, who cares for his own garden with layers of organic mulch and compost made from his household food waste, using no-till methods "means the soil absorbs water faster and encourages keeping water in the soil as a reservoir. This stored moisture in the soil provides another measure of drought resistance during dry spells."

Advocates of surface cultivation are also enthusiastic, but few of them can come up with any real numbers or studies to compare with the yields produced by other methods.

Max Alth, a gardener in New York state, claims in his book *How to Farm Your Backyard the Mulch-Organic Way,* that it takes him only 20 minutes per year to grow 100 pounds of vegetables. He grew his garden ("farm") much the way Ruth Stout did.

In my search for this information, however, I also uncovered some measured surface-cultivation yields in articles written by a man with the delightful name of Ben Easy. Easy's original work was published in the January 1952 and October 1954 issues of the British publication *Mother Earth*, and cites an informal seven-year study done by J. H. L. Chase in England. The comparison of yields in this study was between the produce from a single-dug plot and that of a surface-cultivated plot. As Figure #34 [next page] reveals, some surface-cultivated crops—such as cabbages and cucumbers—actually out performed deeply cultivated crops in both the highest yields and the average yield, even producing a

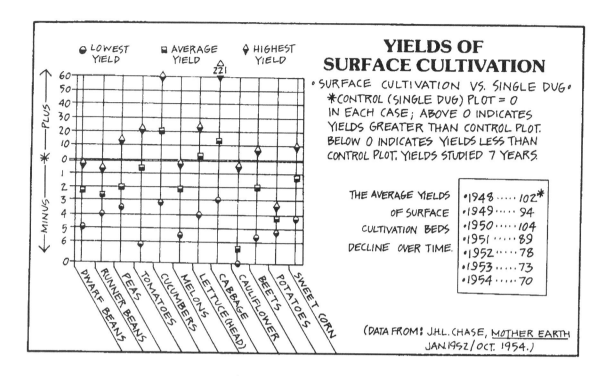

**Figure #34:** It's very hard to find data backing up the claims of no-till gardening. Certainly Ruth Stout never kept records of her days tilling the garden versus her days of no-till, deep-straw beds. The above is one of two examples I know of that provides data about surface cultivation.

2% greater yield the first year. Over time, you'll notice that total yields drop to only 70% of those of the cultivated (single dug) "control" plot. This simply means you again have the choice between vertical or horizontal gardening. You can always add more nutrients via single digging to restart the low-till process or spread out your planting area to compensate for reduced yields.

## A Successful Market Farm Using Surface Cultivation

While most readers will not be growing a market garden, I did discover a fine example of the efficient use of surface cultivation in a commercial setting, which indicates that it can, in fact, be adapted to a larger scale of planting. Most of the following techniques can be adapted to the home garden.

This anecdotal evidence was collected from an organic market garden in England with 12 years of experience in surface cultivation. This garden was not only successful, but "sales were on a competitive basis." More amazingly, the garden was maintained *without the use of any manures*. The book about this remarkable market garden is entitled—at great length—*Intensive Gardening, Using Dutch Lights, Surface Cultivation and Composting for the Commercial Production of Crops, and Introducing a Motion-Study Routine,* 1956 by Rosa Dalziel O'Brien.

Here are some highlights from the book, which help to illustrate how this market garden was put together:

Compost, compost, and more compost. The composting area consisted of "bins" made of bales of straw stacked in the shape of the letter "E," with two "Es" connected in a row to create four bins. The bales of straw were the only significant imported resource. The compost was made with layers of soil, grass clippings, legumes (such as fava beans or vetch, cut and recut from beds not planted in vegetables), and garden waste. The clippings came from grass that was planted in access roads and between rows of plots and mowed when long enough to produce clippings of about three-and-one-half inches. (Be informed that many English gardeners in mild-winter areas mow their grasses all year long.)

Each compost pile, in its straw-bale "bin," consisted of a mass about six feet square and up to five feet high. Every pile began with a small layer of charcoaled wood (a small import or made by burning wood on the property). The purpose of this layer was to "…absorb the impurities from the pile; [it will] serve several successive ones before fresh pieces are needed." [This is an unusual approach, which I'd never encountered before finding this book.] Rosa O'Brien describes a very specific technique for layering of grass clippings, old straw from previous "bin walls," garden waste (thinnings and trimmings, etc.), thin layers of soil, and a very fine dusting of lime—

another import, albeit a small one.
The role of the lime in O'Brien's composting recipe was to make the pile slightly alkaline, in order to enhance existing populations of the free-living, nitrogen-fixing bacteria, *Azotobacter croccoceum*. O'Brien explains: "Our compost heaps include sufficient lime to keep the material alkaline, so that we gain, by bacterial action, about 25% more combined nitrogen than went in with the plant residues, and when we spread the compost on the surface of the soil, the bacteria go on working…[the resulting bacterial action] is quite enough for the crops we grow" (19).

O'Brien also stressed how important it was to use a compost inoculant called Quick Return (Q. R.), made of a special blend of herbs. Each compost pile is pierced by six vertical holes made with a crowbar or broom handle. The diluted inoculant mixture is drained into these holes, after which the pile is capped with two inches of soil. There are five detailed pages devoted to making the piles, and the text refers frequently to the Q. R. solution as an essential part of the process. The solution eliminates the need to turn the piles and can generate finished compost in six to eight weeks in the spring, summer, and fall, or eight to ten weeks in the winter (England). Even though the book was published in 1956, the same (Q. R.) powder is available (as of 2007, when this book was written) from Biocontrol Network in Brentwood, Tennessee or Harmony Farm Supply in Graton, California and other sources.

Other "imports" into this garden included the one-time purchase of so-called "Dutch lights." These are panels of glass built to a standardized form and set up like cold frames or small greenhouses. This was a crucial aspect of the market farm, as the Dutch lights allowed for an earlier harvest and extended the harvest time beyond that of crops grown in an open field. They wouldn't be needed in a home garden if the gardener didn't want to push the natural growth cycle of the vegetable. Also imported, as needed for certain crops, were some seaweed, old domestic soot (from coal-fired chimneys), granite dust, and silver sand.

The main tool for this approach (in addition to hand-pulling weeds or vegetable plants), is a "scrapper," a version of the onion hoe mentioned earlier in this book. [See Figure #20, page 46, in the discussion of corn.] The market gardeners firmly state that: "…preparation for the following crop is one operation with the [scrapper], with the occasional use of a rake, and it should not penetrate the soil more than three-and-one-half inches to four inches for any purpose." The rationale is that, "True aeration of the soil is achieved in undisturbed soil…The scrapper [ED. Known as a scraper, or Dutch hoe, in the US.] is always pulled toward the gardener at a 45-degree angle to the path. The tool is used only to pull, not push soil" (42, 46).

O'Brien mentions that it took about three years for the new market garden to make the transition from harboring obnoxious weeds like docks (*Rumex* spp.), thistles and dandelions to less noxious weeds such as nettle, groundsel [a European weed (*Senecio vulgaris)* with small yellow flowers], and chamomile (probably the low-growing weed (*Matricaria recutita*) not the kind used for tea (*Anthemis nobilus*). It took about the same amount of time for the compost and surface cultivation to enrich the soil, increase its population of beneficial bacteria, and enhance its aeration in a natural way.

With its limited number of imported items and its calculated non-use of manures because the gardeners were vegans ["…(no) animal products or by-products, even egg shells, were included."]—the fact that this market garden was profitable is truly astonishing.

## No-Till Gardening in the Home Garden

Nature builds soil from the top down with all the decomposition of carbonaceous materials such as fallen leaves and decaying grasses, as well as various forms of feces. In addition, there's a lot going on, root-wise, beneath the surface of the soil that helps it improve without cultivation: the activities of roots loosen the soil; the roots of nitrogen-fixing plants add nitrogen; "compost" is created as roots rot at all levels to create decomposed organic matter for the use of other plants and microbes; nutrients cycle from root systems to the foliage of plants (and back to the earth again as leaf, stem, branch, and trunk litter); roots develop tunnels for the movement of

earthworms; root canals help rainwater soak into deeper areas; and root exudates create a microbial "soup" (the rhizosphere) that helps to liberate all manner of nutrients.

Spring apparently causes an unusual hormonal flush in many gardeners, an itch that seems to activate the irresistible impulse to get outdoors and *cultivate, cultivate, cultivate*. As daytime temperatures rise and the earth warms, beginners and gardening veterans alike can't wait to dig in and gently turn over that first delicious rich-smelling spadeful of dark, crumbly loam—a garden's black gold. But wait! Before you reach for that shovel or spade, consider whether cultivation is actually "natural," or even good for the planet.

Our planet's soil contributes ten times more carbon dioxide to the atmosphere than all of humankind's activity; that's 60 *billion* tons of carbon dioxide per year, folks. This odorless gas is produced by the earthy lifestyle of the myriad of creatures that inhabit the soil—microbes, bugs, worms, fungi, and algae—as they breathe, "pass gas," and expire.

Tillage, according to Professor of Biology Tyler Volk of New York University, is a large contributor to the surplus of carbon dioxide and thus to global warming. When soil is stripped of its living cover in order to grow crops, up to one-fourth of its carbon is released, since the carbon dioxide and most of the carbon in the surface layer is seldom replaced. The warming of the bare soil in the sunlight also speeds up the activity and death of soil microorganisms, thus increasing the outgassing.

Gardeners can, in Volk's words, "compensate, by adding compost, plant matter, and manure." But, in reality, most of these sources are really just stored carbon borrowed from other places. Most organic gardeners import large quantities of organic matter and till it into their garden plots or flowerbeds. This well-intentioned tillage, however, contributes, in a minute way, to speeding up the release of carbon dioxide at a time when the planet, is perhaps less capable of reabsorbing the carbon. The rise of one degree (F) or more, already measured in this century may have resulted in part from an extra billion tons of carbon dioxide pumped from the soil into the atmosphere. Currently, the global carbon cycle (in which the rapidly-diminishing rain forests play a big role) may be able to absorb that billion tons, but there's no telling how long it can continue to do so. In your own tiny way, you can minimize the loss of carbon dioxide from your garden, you may want to experiment with no-tillage methods.

So much of what is called organic or environmentally-sound gardening is supposedly based on natural models. For the practice of soil cultivation, the true natural model is, believe it or not, a landslide. Horticulturist John Jeavons maintains that, "Two thousand years ago, the Greeks noticed that plant life thrives in landslides." Jeavons is famous for promoting a method known as "Grow Biointensive®" (GB) a type of raised-bed gardening introduced by the English horticulturist Alan Chadwick who called it "biodynamic/French intensive gardening."

While the aerated soil heaped up at the bottom of a slide may produce lusher crops [where the bigger particles are found lower in the slide (alluvial fan) and the size of the particles graduates to smallest at the top—it is thought that aeration and good drainage favor the growth on the lower portions of the slide], however, landslides just don't happen conveniently in your garden.

The GB method is best known for its use of "double digging," a method of cultivation in which, ideally, all the soil layers, i.e., the natural strata of soil and microorganisms, are kept in relatively the same position they occupy in nature. Soil is moved laterally in such a way that the upper horizons of the soil stay relatively in the same zone as before the digging. The goal is the development of good soil structure through deep soil preparation. Double-digging is used until this occurs. Afterwards two-inch surface cultivation with a Hula Hoe is recommended.

The growing of compost crops to maximize precious biomass resources and add root material to the soil, and the use of compost are also important parts of the GB method.

Most people, when cultivating, are likely to turn the soil, placing the upper soil layers extra deep, accelerating the release of carbon dioxide due to the death of soil life that is buried at a deeper level than where it prefers to live. In the purest form of double-digging, there is no willy-nilly tossing and churning of the soil to two shovels' depth.

This method is done entirely within the soil, resulting in raised beds (due to increased air space in the soil and the incorporation of compost) with sloping sides. It does not in any way resemble the above-ground wooden bins filled with the same soil mixture throughout that is sometimes referred to as a "raised bed."

## Sheet Composting

Sheet composting offers an easy way to improve the garden's soil without strenuous digging. The technique is akin to Ruth Stout's deep-mulching practices. Sheet composting has nothing to do with woven percale or thread counts, but refers to the use of thin layers of compostable material laid out over the soil like a thick mulch. By layering high-carbon wastes with nitrogenous plant refuse, you essentially construct a thin, wide, two-dimensional compost pile. Use a blend of dry brown leaves or woody stems (chipped, chopped, or not); fresh grass clippings; green-manure crops such as buckwheat, vetch, bell beans, and clover; wet kitchen garbage or scraps (no meat scraps, as they'll tend to attract hungry animals and/or smell badly); and green weeds from the garden before their seeds ripen.

Experiment with a ratio of four parts green matter to one to two parts dry dead matter. Water the dry materials as you're layering. At each planting location, make a large planting pocket in the native soil to receive the roots. NOTE: sheet composting doesn't generate enough heat to kill weed seeds, diseases, or pathogens, so after planting, water everything again, then cover the area with five to ten sheets of newspaper, and the newspaper with a weed-free mulch; this should take care of the nasties. The sheet composting slowly helps improve soils from the top down while allowing the gardener to quickly dispose of large quantities of compostable materials and avoid unnatural tillage. The sheet-composted area may require more irrigation during the first year or two if perennials, shrubs, and trees are grown in the uncultivated soil.

## Hay-Bale Culture

Growing annual vegetables on hay bales is a way to rapidly add lots of compostable organic matter to the soil's surface in a short period of time. Here's the recipe:

- Place four or more bales of hay (which has lots of fresh, green nutrients) or straw (with more carbon than green nutrients) in a cluster. Place the bales on their sides, with the end grain facing up, and leave the strings or wire on them.

- Thoroughly soak the cluster of bales.

- Add an inch-thick layer of blood meal or several inches of fresh manure to the top

  (in this case, the exposed end grain) of the bales.

- Add four or more inches of soil to cap off the top of the bales. (All right, I know this requires some digging. But, it's only a one-time effort to start off a remarkable process.)

- Water all the layers again.

- Plant potatoes with plenty of loose straw on top—six inches or more.

- Keep the potatoes and bales moist.

*Et voilá!* The bales will grow potatoes above the ground where no gophers can get to them.

After one season of planting and harvest, the bales will have rotted down quite a bit. Plant them with bush beans, fava beans, other types of beans, or potatoes again.

After the second crop is finished, you may find that the soil is incredibly rich, as the moisture beneath the bales has allowed gophers to till the soil without eating the crop—at least for the first year. The moisture also attracts worms in dry-summer areas, and promotes the growth and health of many helpful microbes, bacteria, nitrogen-fixing algae, etc. After several years of straw-baling crops, you'll be left with a loamy soil and can use any remaining straw to mulch the next crop and add top-down nutrition. Then you may be able to use Ruth Stout's methods to maintain the soil for vegetable growing, or plant perennials, shrubs, or trees and continue with deep mulching or sheet composting.

## Mighty-Good Garden Mounds

Heaping piles of garden "waste" will provide a rich area in which to grow beautiful ornamentals and/or vegetables. This mounding approach is like building a very large, low-heat compost pile where you plant directly into a safe, neutral "cap" of topsoil. Mounds can also be compared to big heaps of "sheet compost."

Start with fresh chunky wood chips mixed with green leaves and twigs from local tree-trimming services. The best wood chips come from hardwoods (and avoid any shredded material that might easily resprout—willows, alders, or various vines). Also avoid plants with foliage that tends to stunt or kill the growth of other plants growing nearby; examples include black walnut (*Juglans nigra*), sagebrush (*Artemisia californica*), and mesquite (*Prosopis glandulosa torreyana)*.

If you can't get a fresh load of green chippings, get mixed sizes of woody chips. Because the chips are so high in carbon, layer or mix them with some manure. Put the chunkiest, largest-sized pieces at the base of the mound to keep the bottom of the pile from settling too much. For all of the subsequent layers, use a blend of smaller chips (1/2" or longer), and materials like: fresh grass clippings; green-manure crops such as buckwheat, vetch, bell beans, and clover; wet kitchen garbage or scraps (no meat); and green weeds from the garden. Experiment with a ratio of at least four parts green matter to one part wood chips. The more nitrogen (green material) you add, the faster the mound will decompose and the greater the nitrogen supply for the growing plants will be. Pile this mixture of high-carbon and nitrogen-rich materials at least 1/3 higher than your projected final soil level. Depending upon the types of materials used, the mound may settle as much as 30–50 percent.

Make the soil cap at least four inches thick—the thicker, the better. (Again a one-time use of digging.) I usually try to dig this soil from the ground adjacent to the mound and to curve the dug-out area like a small, serpentine stream bed, so that as I add soil to the top of the mound, the swale (drainage area) gets deeper at the same time. This doubles the depth of the swale as the mound is built, providing better drainage during heavy rains. Using native soil gathered from around the garden will usually ensure good drainage, a normal soil temperature, and good initial growth. The combination of materials used in the mound acts like a slow-heating compost pile. The root hairs of mound-planted specimens will follow right behind the warm decomposition and won't grow into areas in the core of the pile that are too hot due to hot composting activity. The plants also get some additional benefit from the heat the mound generates, much like bottom-heated greenhouse flats.

I use the mounds to plant perennials, herbs, sub-shrubs like lavender, and other shrubs and trees. If I plant in the fall (just as the rains begin in my Mediterranean climate), the roots grow all winter and no irrigation is needed in the summer from that point on. Be reminded that I live in a relatively cool, coastal environment. In hotter summer areas with cold winters, spring planting and some periodic irrigation may be required.

Eventually, the whole mound settles to a lower level, the chips fully decompose, the shrubs root into the native soil, and you're left with a wonderful, curvilinear feature in the landscape. Mounded plantings then just need a seasonal weeding and perhaps an occasional application of newspaper layered with weed-free mulch.

# CHAPTER 8

# Fruit Tree Roots

W hile Eve was picking that juicy apple, she probably wasn't thinking about tree roots.

If the apple tree were more than metaphorical, however, you can be sure that while Eve and her hubby were sampling forbidden fruit and getting in trouble with the Boss, its roots went on serenely hunting for food, water, and healthy aerobic soil.

The best book on the topic of fruit tree roots is *The Root System of Fruit Plants* compiled by a Russian named V. A. Kolesnikov. (As with most scientific papers, only initials are used for all

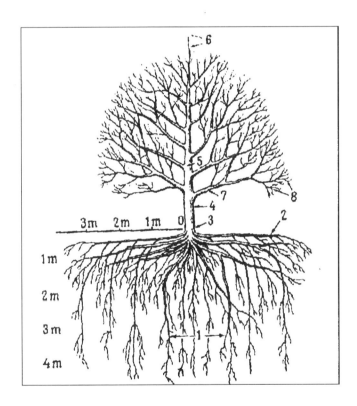

**Figure #35:** This is Kolesnikov's view of the basic components of a typical fruit tree. Notice that there is no true taproot. The root system is fibrous with horizontal and vertical roots. #1 = vertical roots, 2 = horizontal roots, 3 = root collar, 4 = trunk, 5 = stem, 6 = central leader, 7 = main branches and 8 = laterals. Why this illustration doesn't show a root zone significantly wider than the foliage is beyond my comprehension.

All illustrations in this chapter used by permission of The University Press of the Pacific. From: *The Root System of Fruit Trees*, V. A. Kolsnikov, the 2003 version of the original 1971 edition. (Unless another source is indicated.)

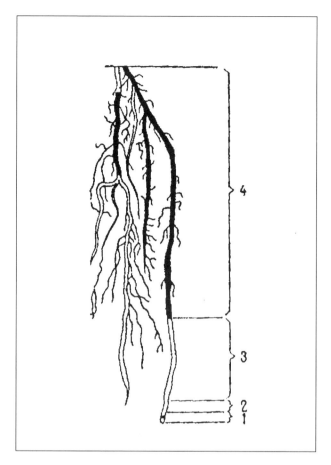

**Figure #36:** 1 = the root cap, 2 = cell division and cell elongation, 3 = absorption zone, 4 = conducting zone.

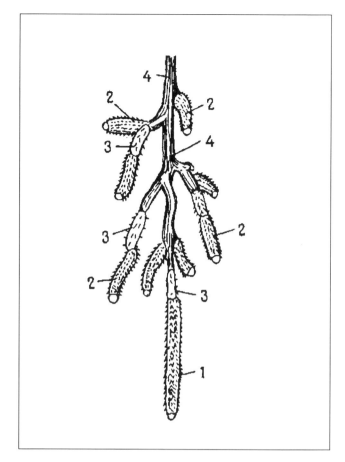

**Figure #37:** The fibrous roots of an apple tree. 1 = growing or axial roots, 2 = absorbing or active roots, 3 = intermediate roots, 4 = conducting roots.

but the surname.) Kolesnikov's scientific papers appeared from 1924 until 1968, indicating that the USSR regime certainly valued his distinctive research.

Kolesnikov's primary *modus operandi* for studying roots was one he called "The Skeleton Method." As with Professor Weaver's studies, this method entailed precise excavation of the roots. In imitation of archaeological techniques, shovels were used first, followed by scoops, and eventually, brushes. This approach preserved more fine root hairs during excavation than the most common practice of using water

forced from a hose, or even more than simple washing of the roots. In his book, Kolesnikov goes into excruciating detail about how to excavate and document the entire root system of a fruit tree. The process seems so tedious and backbreaking that it's a wonder anyone even tried it. In spite of this, Kolesnikov mentions seven other researchers who also used the Skeleton Method.

Thanks to Kolesnikov's work we have some amazing drawings of how a fruit tree's roots really grow; not surprisingly, these look much like Weaver's illustrations of native root systems from Chapter 4.

Kolesnikov begins his book by categorizing the main elements of a tree [See Figure #35], and spends some time explaining the importance of roots. He remarks, for example, that shoots of a tobacco plant grafted onto the roots of a tomato plant do not produce foul-smelling addictive smoking material, but rather, normal-looking nicotine-free leaves. Conversely, grafting tomato cuttings (scions) onto the roots of a tobacco plant produces tomato leaves with four percent nicotine. In another example, albino sunflowers were grafted to green sunflowers and produced only green sunflowers. Both of these examples seem to indicate that the roots of a plant play a role in the formation of chlorophyll and other substances in the leaves.

What Kolesnikov calls the *absorbing* or *feeding* roots are those where the root hairs are found. [See Figures #36 and #37 to see the parts of the feeding roots as drawn in his book.] The absorbing roots are very active; at the peak of a tree's root growth they account for more than 90% of all its roots. Thus, Kolesnikov's comment in the introduction of the book that, "at the end of the first year's growth, an apple tree can incorporate 17,000,000 root hairs, with a total length of 6,562 feet." The root hairs can increase a tree root's "feeding" area by two to ten times!

Kolesnikov proposes that growing roots are responsible for the amounts of humus, microbes, and worms that accumulate in the soil around each tree. Other scientists have adopted the theory that roots are opportunistic and follow the holes left by burrowing worms to feed on the humus. It's indisputable that dying root hairs on the roots growing up into the duff provide lots of compostable material. Kolesnikov calls this process *root shedding* and notes that it happens throughout the root system when parts of the root mass die naturally and leave behind organic material for the soil's flora and fauna to digest. In that sense, the roots are developing their own form of humus.

While it was a scientist with the pleasant-sounding name of Du Hamel du Monceau who (in 1758) was the first to write about the "self-thinning" of roots (that is, the process of new absorbing rootlets forming after a small root's demise), it was Kolesnikov who did the most extensive work on this subject. He found that the death of absorbing roots equaled two tons per two-and-one-half acres of a 25-year-old spruce (*Picea* spp.) forest. In many trees, he discovered, the roots provide as much as two tons of compostable material for every two-and-one-half acres and are therefore responsible for the increase of humus in all layers of the soil. Cultivation increases the formation of laterals in the upper soil horizon where the most feeding roots are found.

## ❧ PRACTICAL TIPS FOR GARDENERS

Roots find conditions the most cushy and convenient near the surface of the soil and will do almost anything to live there. Since they can find their way through openings in hardpan or less fertile soil, there's no real reason to try to break up a hardpan if there is a reasonable amount of soil above it. If the drainage is poor and water is likely to flood the roots, and if there is some decent soil beneath the hardpan, then breaking through will probably be required. You might prefer to plant your tree on a mound, so that the crown of the root system doesn't get root rot. [See Chapter 15.] Remember that the best place for water, fertilizer, compost, and mulch is beyond the foliar drip line.

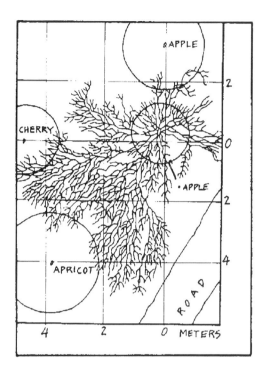

**Figure #38:** This apple tree's roots extend far beyond the circle representing the canopy of the tree and avoid the compacted soil of the roadway. (Scale is in meters.)

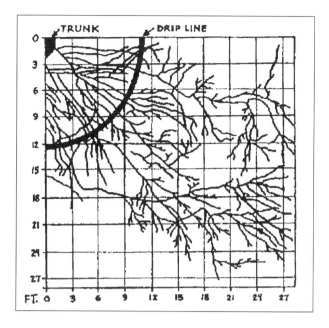

**Figure #39:** This shows how far a 45-year-old apple tree's roots spread beyond the dripline, represented by the the arc in the upper left side of the illustration. (The scale is shown in square feet.)

Some tree roots tend to avoid each other. In Figure #38 the circular lines represent the canopy of the trees. Here we see that the apple tree's root system tends to grow away or head downwards as it approaches that of another apple tree. (A puzzling observation since apple trees seem to thrive in orchards.) The roots also naturally avoid compacted soil as found on the nearby road. Figure #39 also shows the ratio of the root to the canopy for a 45-year-old apple tree.

Kolesnikov's conclusion is that fruit-tree roots grow one-and-one-half to two and even three times the width of the foliage above them. More amazingly, he states that this ratio is maintained throughout the life of the tree, *regardless of the rootstock, species, and soil* (my emphasis added). This is clearly seen in the apple trees depicted in Figures #39 and #40. As Figure #41 shows, each

type of fruit tree maintains a slightly different ratio of root mass to canopy.

The relationship of the width of a tree's root-mass to the amount of moisture it should receive is critical. As I've written in previous chapters, applying water near the trunk is wasteful in any climate. In a climate that routinely experiences short droughts of a month or so up to six months (as in parts of the Southwest), drip irrigation is the most efficient way to distribute water to an entire root system.

The climate, however, need not be arid for trees to benefit from drip irrigation. In a study of established pecan trees in humid Georgia (USA), trees with added drip irrigation showed a 51% increase in yields.

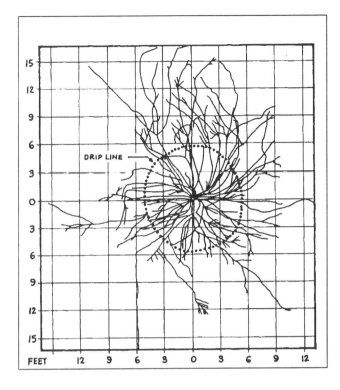

**Figure #40:** The full root system of a ten-year-old apple tree. (The scale is in one-foot squares.)

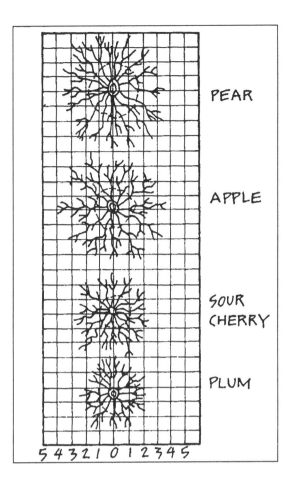

PEAR

APPLE

SOUR CHERRY

PLUM

5 4 3 2 1 0 1 2 3 4 5

**Figure #41:** All trees are not created equal. This illustration, in one-meter squares, shows how different root systems spread.

Paul Vossen, the University of California Cooperative Extension tree-crops farm advisor for Sonoma County, simultaneously tested many possible kinds of irrigation systems for fruit trees (including drip irrigation). He has evaluated the systems' performances for both home and commercial plantings, and demonstrated to both home gardeners and farmers how various types of hardware might work for them. The drip-irrigated trees at the Germone Demonstration Orchard showed some impressive results. Paul's records over six years show that, "With some of our peach trees, we're getting 14 tons per acre in the third year, while local growers are only getting seven tons, and on three-year-old apple trees we're getting 20 tons per acre, while established, unirrigated, and mature apple orchards are only getting 13 tons per acre." (These impressive increases in yields translate to an increase in abundant foliage and plentiful bloom produced on ornamental trees when appropriate irrigation is used.)

In my orchard, I prefer frequent watering with small amounts of water, sort of like "topping off the tank." After the Northern California winter rains are over and the soil has reached an ideal moisture level, not too wet and not too dry, my goal is to replace, as often as daily, exactly the amount of moisture lost due to evaporation from the soil and transpiration from the plant's leaves (called the evapotranspiration rate, or ET), plus an amount that represents enough extra water for gorgeous growth. *Arboriculture* (by Richard W. Harris, James R. Clark, and Nelca P. Matheny), the preeminent text on growing and caring for trees, also recommends frequent irrigation: "In contrast to other systems, drip irrigation must be frequent; waterings should occur daily or every two days during the main growing season… the amount of water applied should equal water lost through evapotranspiration."

## ℝ PRACTICAL TIPS FOR GARDENERS

Where should you water your fruit trees? Throughout the entire root zone! [See Figure #42 for some ideas for laying out a drip line beneath and around trees and shrubs.] I'm a proponent of in-line drip tubing, in which the emitters are built inside the tubing at regular intervals— nothing to break off! [See Appendix #1 for more details.] With the regular and appropriate placing of emitters, you can irrigate an entire root zone some four to 12 inches below the surface of the soil. The dry spots on the soil surface are deceptive, since the underground bulging of the wet zone provides continuous irrigation below ground [See Figure #43.] As a tree grows, you can add lengths of in-line tubing to keep the new roots happy. Or, consider laying the in-line tubing over the whole zone before you plant, as shown in Figure #44. Another option is laying out the tubing around existing trees and

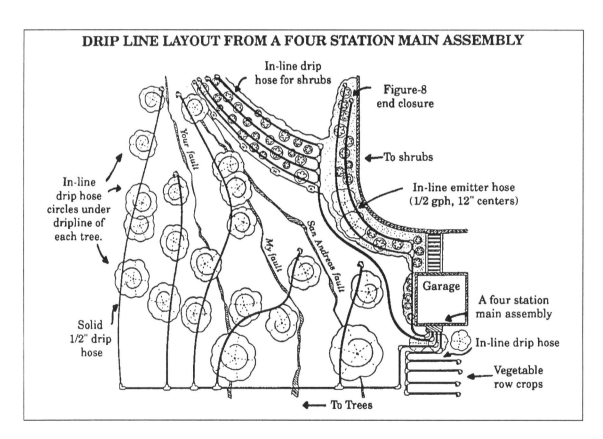

**Figure #42:** Several options for drip irrigations for your garden: parallel lines work for closely-planted shrubs and for the vegetable garden. Circles of tubing can be placed beneath the dripline of each tree. The underground wet spot will spread beyond the dripline of the tree's foliage. (From: *Drip Irrigation, For Every Landscape and All Climates.*)

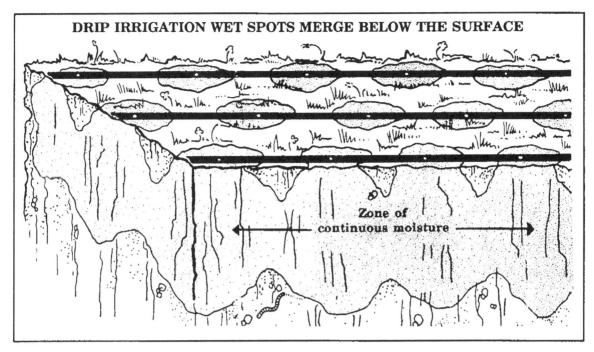

**Figure #43:** While there are dry areas on the surface of the ground, the moist areas watered by the drip emitter will merge together beneath the surface of the soil to irrigate the full root zone. This only works if you use the right spacing, based upon your soil. Sandy soil requires 12-inch spacings and heavy clay might be 24-inch spacings.

(From: *Drip Irrigation, For Every Landscape and All Climates.*)

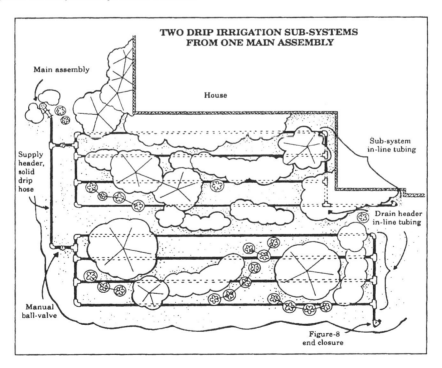

**Figure #44:** An example of rows of tubing placed prior to planting to irrigate the entire root zone of the prospective fruit trees and shrubs. (From: *Drip Irrigation, For Every Landscape and All Climates.*)

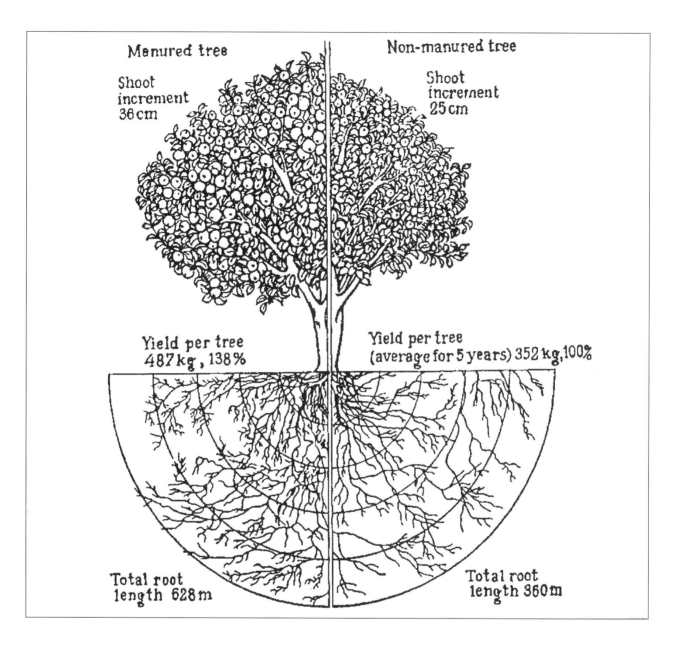

**Figure #45:** This is Kolesnikov's illustration of what fertilizer, manure in this case, can do to influence the yields of a 'Golden Winter Parmen' apple tree. The roots are not much wider than the dripline on the manured side of the tree. This is perhaps because placing the manure under the foliage concentrates the roots at the source of the fertility. The best place for irrigation is at the dripline in this example.

shrubbery, as shown in Figure #42. Figure #45 shows how important it can be to irrigate and fertilize.

How much water should you use? As mentioned above, the loss of water each day is measured by its evapotranspiration rate, or ET. The chart in Figure #46 shows how to translate inches per month into gallons per day (GPD). Call your local Cooperative Extension to get the average ET rate for each month in which irrigation

# Daily Water Use (In Gallons per Day)
## BASED ON VARIOUS EVAPOTRANSPIRATION RATES

| Square Feet of Plant Cover | ET Rate (in inches/month) | | | | | | | | | |
|---|---|---|---|---|---|---|---|---|---|---|
| | 1" | 2" | 3" | 4" | 5" | 6" | 7" | 8" | 9" | 10" |
| 1 sq. ft. | 0.0187 | 0.0374 | 0.062 | 0.083 | 0.104 | 0.125 | 0.145 | 0.166 | 0.187 | 0.208 |
| 4 sq. ft. | 0.075 | 0.15 | 0.248 | 0.332 | 0.416 | 0.5 | 0.58 | 0.664 | 0.75 | 0.832 |
| 10 sq. ft. | 0.187 | 0.374 | 0.62 | 0.83 | 1.04 | 1.25 | 1.45 | 1.66 | 1.87 | 2.08 |
| 75 sq. ft. | 1.403 | 2.805 | 4.65 | 6.225 | 7.8 | 9.4 | 10.875 | 12.45 | 14. | 15.6 |
| 100 sq. ft. | 1.87 | 3.74 | 6.2 | 8.3 | 10.4 | 12.5 | 14.5 | 16.6 | 18.7 | 20.8 |
| 200 sq. ft. | 3.74 | 7.480 | 12.4 | 16.6 | 20.8 | 25. | 29. | 33.2 | 37 4 | 41.6 |
| 300 sq.ft. | 5.61 | 11.22 | 18.6 | 24.9 | 32.2 | 37.5 | 43.5 | 49.8 | 56.1 | 62.4 |
| 1 acre solid cover | 8.15 | 1629 | 2701 | 3615 | 4530 | 5445 | 6316 | 7231 | 8146 | 9060 |

**Figure #46:** This evapotranspiration (ET) chart will help you calculate how much water to apply as determined by your local climate. An ET of 10 inches is something like that of Palm Springs in the summer. To be accurate, contact your local Cooperative Extension agent for the ET rate for the seasonal or monthly average. If they don't know what you're asking for, fire them.

(From: *Drip Irrigation, For Every Landscape and All Climates*.)

is needed. Use the chart to figure out how many GPD are need to replenish the water lost via evapotranspiration. You can replace this on a daily basis or at a weekly interval—just multiple the daily amount by seven. Knowing how many emitters you have in a planting zone will allow you to do the math to tell how long to leave a drip system on. For example: if you have 100 square feet of planting and 25 one-gallon-per hour emitters and the ET rate is six inches per month, you need to run the system about one-half hour per day. (100 sq. ft. = 12.5 gallons divided by 25 one-gallon-per-hour emitters = one half-hour.) or about three-and-one-half hours per week.

# CHAPTER 9

# Native & Ornamental Trees

On February 14, 1986, I was forced to stand up my Valentine's Day date, having been effectively trapped on my property by trees blown over by 90-mile-per-hour winds. More than a dozen three-foot-diameter Douglas fir trees (*Pseudotsuga menziesii*) lay across the road that separated me from my loved one. While this was both socially and environmentally unfortunate, I did learn a lot, watching in horrified fascination as the winds slowly rocked these stately and enormous trees to the ground. My first revelation, on venturing out after the storm, was that, in spite of my assumption that Douglas firs grew a taproot, the roots revealed on these toppled trees were relatively shallow and wide. To my amazement, most of the 50- to 100-foot-long trees lying around in the carnage had generated their roots only in the upper two feet of the soil. Moreover, in the area of their root mass where one would expect a taproot, I could find only a clump of large roots splayed out like some kind of flattened brown octopus.

The pattern of the Douglas firs' secondary roots, the laterals, was equally fascinating. How, I asked myself, did these trees stand up in such a shallow soil above a barrier of clay and stone? Part of the story was revealed in the mass of roots that popped out of the ground as the trees fell. As much as eight to ten feet away from the trunk, some of the torn and shattered roots were a husky two to six inches in diameter. How much further

away from the tree would I have to trace them before they would dwindle to spindly rootlets?

While it didn't do much for my love life, this stormy episode renewed my curiosity about root-growth patterns. So, back I went (once the road was clear) to the University of California Agricultural Libraries at Berkeley and Davis, for more research on the subject.

## Root Types

As described before, many of the trees we grow in our yards (and most fruit trees) have roots that are fibrous in the sense that after their taproot—if it had one—fades away, the rest of the roots originate from a small area at the base of the trunk, sometimes called the crown of the root system. Many trees grow major horizontal roots called *laterals* (not to be confused with the very small lateral roots that comprise the root hairs) most of which will be found within the top one to three feet of the soil. Numerous vertical roots, called *sinkers*, will descend anywhere along the length of the laterals. Some roots also grow more at an oblique angle than a vertical one; it's these roots that can extend to reach further than the perimeter of the dripline.

What allows a native tree to stand up to formidable winds is the width of the laterals and the anchorage of its root system. This may be composed of several factors: deep penetration (up to three to four feet), coupled with the roots "holding onto" rocks by growing tightly around them; an extensive surface root system; root grafting (where roots cross and are fused together as their diameters expand); and the sheer strength of the oblique and vertical sinker roots that can ramify so much soil that the weight of the soil with roots below ground can be six to sixteen times the weight of the tree aboveground. Some say this root mass can equal eight to twelve times the total weight of the tree. If you don't believe the power of a sinker root, find a tree that has blown over and try to pull even a one- to two-inch sinker root out of the ground. It will put up a surprising fight.

## The Root of the Matter

I soon discovered that in a good soil, a tree's roots will often grow to occupy an underground area wider than its dripline. If led by available moisture and nutrition, a tree may tunnel its roots through soil space ranging from an area one-half wider than the dripline to as much as three times further. In special cases, tree roots may ramify much more than anyone would imagine [See Figure #47.] If you add a subsoil barrier such as rock, bedrock, or caliche (hardpan), a tree's roots will wander even further beyond the canopy area in search of food. A deep sandy soil offers little resistance to growing roots and allows for root exploration of three or more times the width of the tree. And, like a gardener struggling to dig heavy clay soil, roots don't like clay either and

don't make much headway through it, perhaps only one-half the width of the dripline. In Figure #48, you'll notice that the canopy of this walnut tree measures about ten to fifteen feet in diameter, yet its primarily shallow roots (most of its lateral roots grow in the top one to two feet of the soil) extended for fifty-one feet before the researchers finally gave up—and the root still measured one-half inch in diameter. (Those fainthearts obviously never met Professor Weaver!)

Some more examples include:

- Poplar (*Populus generosa*) can ramify 77% of its roots beyond the dripline.

- Another study found that 35% of poplar trees grew roots greater than two times the distance from the trunk to the edge of the foliage.

- Colorado blue spruce (*Picea pungens* 'Glauca') grow 60% of their feeding roots beyond the dripline.

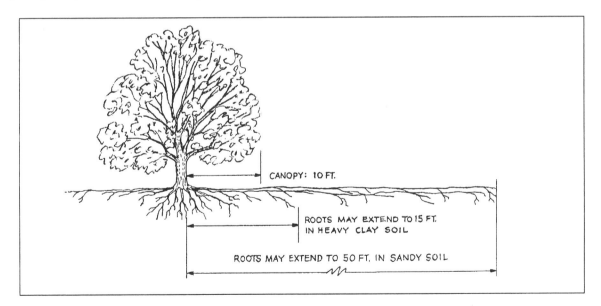

**Figure #47:** Trees' roots commonly grow one-half wider than the dripline (canopy), and occasionally to as much as three to five times further.

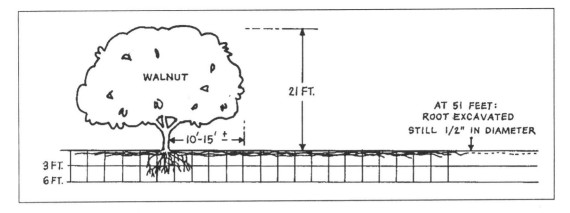

**Figure #48:** Because this walnut grew above a hardpan some three to six feet deep, there is no clear taproot. As with most trees, the majority of the roots are in the upper eighteen inches of the soil.

- White green ash (*Fraxinus pennsylvanica*) grew roots which were 1.68 times the radius of the dripline.

- In one study, the glorious magnolia (*Magnolia grandiflora*), had grown roots 3.77 times wider than the dripline.

- Sugar maples (*Acer saccharum*) produced roots that have been found spreading underground 30 feet beyond branch tips.

- The roots of honey locusts (*Gleditsia triacanthos)* can reach nearly three times beyond the dripline of the foliage.

Another approach to root study is to figure the ratio of the height of a tree relative to its extent of rooting. Figure #49 shows that young trees grow a proportionally wider spread (1.4 times the height) as compared to older and mature trees with roots only 40% wider than the height of the tree. This study (*Natural Root Forms of Western Conifers*, by S. Eis, in the 1978 proceedings of the Root Form of Planted Trees Symposium) was done with Douglas fir trees (*Pseudotsuga menziesii*), which grow in a taller, more pyramidal shape than oaks or other deciduous trees. I'm always being reminded that nature doesn't always follow the "rules" or read my books.

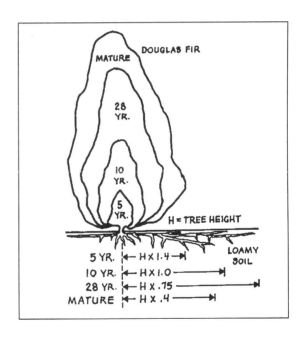

**Figure #49:** This is a formal chart based on the height and age of the tree shown. As the tree gets older, the proportion of root width to height gets narrower.

The point about my presenting these root drawings of trees is that the water, fertilizer, and mulch you apply should extend at least out to the dripline of a tree, bush, shrub, etc.—and preferably further. A doughnut-shaped area of moisture and amendments created beyond the dripline will conserve mulch, fertilizer and labor, as well as encouraging the roots to grow away from the trunk. This lateral growth, combined with "sinker" (vertical and oblique) roots, helps protect the tree from toppling by wind. In fact, with established trees there is absolutely no need for water, fertilizer or mulch at the base of the tree. (In a British study done using radioactive isotopes, the roots four-and one-half feet away from the trunk of a ten-year-old apple tree absorbed slightly less than 10% of the water and nutrients absorbed by the entire root system.) With younger trees, the absorbing zone will be smaller. The base area, say up to several feet, depending on the age of the tree, can be kept clean of weeds with a bit of surface cultivation, mulch, or hand weeding.

As mentioned several times in this book, growing trees from seed provides them with the healthiest root system. The roots will find their natural place in the soil. Patience furthers!

Another item to consider is the use of herbicides and soil-injected insecticides. With roots growing so far from the tree's foliage, it's possible that the use of soil-altering substances in your own yard could accidentally harm or otherwise affect trees belonging to the folks next door.

## ℞ PRACTICAL TIPS FOR GARDENERS

This section of the book contains perhaps some of the most important information for readers, since it illustrates that most gardeners put their water, compost, mulch and fertilizers in all the wrong places. Many other plants grow in the root-to-canopy ratios detailed above, yet many gardeners apply most of their fertilizer and mulch right around the base or trunk of their trees, shrubs, or even herbaceous plants. (Mulch at the base of a tree in a lawn is helpful in preventing "mower burn," which is a result of skimming the ground too close to the upper roots at the root crown, but mulch that is piled high around a trunk may encourage root rot or crown rot.) At least some commercial fertilizers, such as Ross Fertilizer Stakes®, recommend placing their product at the dripline of a tree—further would be better. (Even after all these years, I have to remind myself to pee at or beyond a tree's dripline, rather than at its trunk—unless I need the tree to hide behind!)

## Some More Root Factoids:

Interestingly, some trees, such as many Californian and western oak trees, begin growing with a taproot, which then naturally atrophies. When the young seedling of a blue oak is a mere three inches high, the taproot may already extend forty inches into the soil. After a number of years, the taproot withers, to be replaced by heart roots (which angle down from the base of the tree) and many laterals, with sinker roots. In spruce (*Picea* spp.), hemlock (*Tsuga* spp.), and cedar (*Cedrus* spp.) trees, in less than eight years, the lateral and oblique roots take over the role of support and the taproot declines. After the taproot atrophies, the new root system grows more horizontal and oblique roots and resembles a fibrous root system.

While one rule of limb has been that a tree's roots are one-and-one-half to three times wider than the foliage, other investigators estimate an irregular root pattern four to seven times the crown area; and, still other researchers maintain that the root extension can be four to eight times wider than the dripline of the tree, but only under certain conditions.

Whatever the details, the point is that roots travel much farther beyond the dripline than most gardeners expect.

A study of Sitka spruce tree roots (*Picea sitchensis*) growing on a layer of two types of peaty soils in England produced a number of interesting conclusions:

- Lateral root spread increased with deeper soil when going from waterlogged soil to drier soil.

- The spruces' root depth was between 12 to 18 inches. The mean depth of the roots was 16.5 inches.

- The rooting depth decreased as the diameter of the root "plate" (their term for what I referred to as a flattened brown octopus) increased.

- Roots did not grow more or larger on the side of the tree receiving the most wind.

- The trees in shallow soils showed root grafting that linked two or more trees together. This is a little-understood phenomenon. Some studies show that little or no transfer of nutrients occurs between the trees, while others show a small migration of nutrients between trees. Most studies suggest that the roots happened to grow over each other and fused as they grew in diameter. The one common observation is that a single tree in a grove of trees connected by root grafts is less likely to blow over.

## Seedling Trees

Much of the research on growing native trees has been done by timber companies trying to figure out how to replace clear-cut areas with trees that will grow quickly, with straight trunks and resistance to wind and snow loads. For example, a study by Annie Plourde, at Cornelia Krause Université du Québec à Chicoutimi, Département des Sciences Fondamentales, states: "…It has been shown that the current methods of seedling production and planting cause significant deformations of root systems of Jack pine, such as winding of the roots, poor distribution, no taproot and shallow rooting, and these defects may affect more than 90% of planted Jack pine. In contrast, these types of deformation and the development of poor root systems are not known [in the case of] naturally regenerated [from seed] Jack pines." See Figure #50 for the comparison of young saplings.

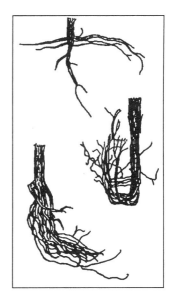

**Figure #50:** Unlike the other two seedlings shown, the upper seedling was planted without the use of a reforestation tool.

## ✿ PRACTICAL TIPS FOR GARDENERS

Big surprise: trees don't naturally grow in pots in the woods. If planted as seeds, they will naturally create a healthy wide-ranging root system without anybody's help. As gardeners, we tend to want it all—now. Sadly, the size of a plant in a container—rather than the condition of its root system—is often the reason it gets purchased by people who are uninformed about how trees actually develop. Large container plants offered commercially in ordinary plastic pots for transplanting into a yard or garden will frequently have developed overgrown roots in an attempt to support the abundant visible growth. With larger plants, this usually means that the cramped root system is "pot-bound"—the roots have circled dizzily inside the container in search of space and nourishment and have tangled themselves into a near-solid mass. Avoid these plants like the plague. Demand that your nursery acquire plants with a young, healthy root system.

## Goin' Tubular

When you buy any normally taproot-growing plant that is potted or in any another commercial plant-buying option, you've basically kissed the taproot goodbye. Patience favors those that want the most fully-rooted plants (whether for a long-term windbreak or as a present to the great-grandchildren), because seed-grown trees are the only plants that will produce a healthy, unhindered root system. Once seed-grown babies start taking off, they can actually outgrow trees transplanted from containers. Examples of taprooted trees include most nut trees, such as walnut, pecan, hickory, and butternut. (Neither almond trees, which belong to the stone-fruit family, nor hazelnut trees grow a taproot.) Many desert trees, shrubs, and perennials grow taproots for survival. Other examples of taprooted trees include many oaks (*Quercus* spp., especially Eastern U.S. oaks), redbud (*Cercis* spp.), Douglas fir (*Pseudotsuga menziesii*), grand fir (*Abies grandis*), and incense cedar (*Libocedrus decurrens*).

A great deal of research has centered on designing containers that air-prune the roots of tree stock. Again, most of this research has been done by the forestry industry, because many of the seedlings valued for reforestation and lumber "plantations" are varieties that primarily grow taproots, rather than fibrous root systems. The industry's goal has been to stimulate a fibrous root system (which is contrary to the plants' nature), in order to lessen transplant death. To reduce the effort of transplanting in wild and rough terrain, the forestry industry is interested in making seedling containers as small and as lightweight as possible. Following these guidelines, they have developed a variety of tube-shaped containers—generally long, narrow, lightweight, and with one or more air-pruning holes (referred to in industry jargon as "root egress openings") in the bottom. [See Figure #51.]

The best way to reduce circling and stimulate fibrous-root growth in a tree is to prune its roots, and an effective way to do this without pulling the tree out of its container is "air pruning." Once a new root-tip hits the air at the bottom of the tube, it dries up and dies, and this stimulates side-shoots further up on the root. In pursuit of the perfect small lightweight container that will air-prune nursery tree stock, researchers have developed a variety of tube-like shapes. The criteria for a good tube container include: easy removal of the seedling; solid walls so roots don't grow from one container to another; ribbed sides to prevent circling and to direct roots downward; sufficient soil volume to support healthy tree-root growth; and an opening at the bottom to air-prune taproots. Richard Tinus and Stephen McDonald, in their paper *How to Grow Tree Seedlings in Containers in Greenhouses*, recommend that the ratio of a tube's depth to its diameter be as much as ten to one. Other researchers recommend that tube containers should be no more than six inches deep.

A tube-grown tree is usually quite short when you buy it—an advantage, not a handicap. I have been very enthusiastic about tube-grown trees and shrubs since the mid-1980s, especially after experiences like the following; back then, some clients of mine showed me several very root-bound Monterey cypress (*Cupressus macrocarpa*) trees, which they had originally purchased in 15-gallon cans. ("Only $95; what a great deal! And they're over six feet tall!") These trees-in-bondage had been held for another several years in the cans and then planted (not by me!) on the most sheltered side of the house. Nevertheless, winds were soon toppling the 15-gallon "bargains." Even posts, ropes, and cables couldn't keep the trees upright, so my clients cut the tops out at six to seven feet, leaving some truly ugly trees that provided neither shelter nor privacy.

Along the windiest side of the house (up to 90–mph blasts in the worst winters), I planted a rambling hedgerow of temporary "nurse" plants, using a shrub called myoporum (*Myoporum laetum*). [I wanted to plant tube-grown Monterey cypress seedlings, but none were obtainable.] Instead, I obtained some smallish cypress plants (eight to twelve inches tall) in one-gallon cans. These were planted in and around the myoporums so the wind was buffered and no staking was required. In my experience, trees staked at transplanting can run the risk of not outgrowing the stake. Within four years these unstaked seedlings had reached eight–ten feet —taller than the $95 "specials." I'm convinced that four eight-inch-tall tube-grown trees, at one-fourth of the cost, would have done even better.

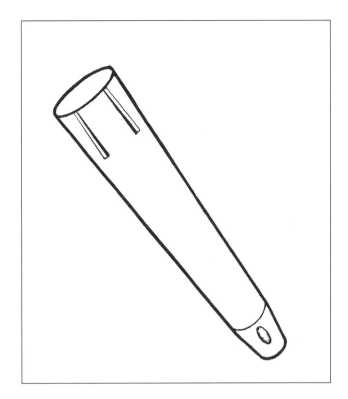

**Figure #51:** A simple device, but it really helps produce more lateral rootlets for less transplant shock. This is one of many shapes.

(The availability of tube-grown stock is still a problem—see "Searching for Sources of Tube-Grown Plants" in Appendix #3. Also ask your local Farm Advisor or Department of Forestry about tree seedlings grown for revegetation projects.)

In summary, according to Richard Harris, professor of Environmental Horticulture at the University of California at Davis, "…The smaller the plant when transplanted into the landscape, the better will be its relationship to the environment."

## Roots, Grow Up!

Surprisingly, many of a tree's feeding roots grow up, not down. In a paper published in *The Landscape Below Ground,* Professor Thomas

O. Perry states, "Most of a tree's feeder roots range in diameter [from that] of a lead pencil to the size of a hair. These smaller roots…grow upward into the surface inches of soil and the litter [duff] layer" They're looking for moisture *and* nutrients.

The significance of the importance of the most aerobic layers of the soil and duff can be seen in a simple study using potted tree seedlings. In Figure #52, you can see the dramatic difference in growth between roots ramifying three soil depths. Whether you're talking native or ornamental, the top two inches of soil are vital to a tree's health.

This is also the favorite soil zone for any plant in a humid climate with periodic rains. [Information on the effects of drought on trees comes a bit later.] Take away that top two inches by planting it to lawn, raking the duff up for "a cleaner look", or allowing so much foot traffic that the roots are exposed, and you have a horticultural disaster in the making.

About 90% of the total fine-root biomass of Pacific Northwest forest trees may grow in the upper one-and-one-half feet of the soil. Figure #3, in the chapter on Humus & Mulch, reveals how and why those top few inches are so

**Figure #52:** I think this is the most revealing illustration in this book as it shows how influential the shallow zones of soil are to the feeding roots of a young tree. The seedlings in the pot on the left are growing in soil gathered from the top two inches of forest soil in a specific growing area. The middle pot contains soil gathered from a zone two to four inches deep in the same patch of ground. The seedlings on the right are growing in subsoil gathered from beneath the topsoil. The pot on the left outperformed all other seedlings by more than 50%.

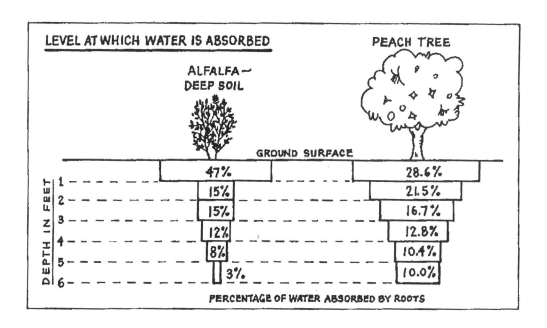

**Figure #53:** Alfalfa can grow roots to dozens of feet deep. Yet, like the taller peach tree, shown here, most of the water absorbed comes from the top foot of soil. With water, nutrients follow.

important. The aerobic-loving soil life needs to breathe. The deeper you go, the less aerobic you get, and the number of beneficial soil flora will rapidly diminish.

Studies done with agricultural plants provide a lot of useful information. Alfalfa and peach trees get most of their moisture (along with nutrients) from the top one to two feet of the soil. [See Figure #53.] Figure #54 details the depth at which various plants and trees "gather" most of their water. Figure #55 shows the nutrient levels of absorption for a black walnut tree, two alfalfa plants, and Taiwan hardwoods.

Figures #53, #54, and #55 are of economic crops—fruit trees, alfalfa, cotton, and corn. I have been unable to find similar data for the depth at which native trees absorb moisture, except for the 30-year-old mesquite tree example in Figure #56. [See page 93.] There seems to be little incentive

to do such research with noneconomic or merely ornamental trees. I am making an unsubstantiated assumption that ornamental trees will have similar absorption patterns as fruit trees because they are planted in about the same way, which means that the taproot is usually destroyed in the process.

In some special cases, roots will grow back up from rather deep placement in the soil to range within two inches of the surface. Figure #56 shows a small 30-year-old mesquite tree (*Prosopis glandulosa*), 25 inches tall and 35 inches wide, growing in southern New Mexico, its roots penetrating more than 18 feet into the soil. The researchers concluded that, along with generating deep roots to gather moisture from heavy rains seeping far underground, mesquite roots "grow…upward, to utilize minor precipitation events that only wet the soil to a depth of a few centimeters (about three-quarters

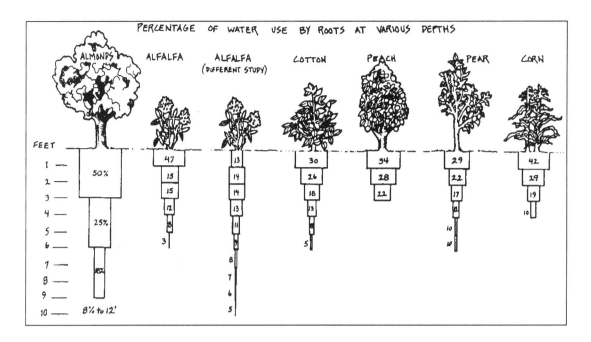

**Figure #54:** The height of a plant does not represent the depth of the root system. (The relative width of the root system isn't shown in this illustration.) Again, the top one to two feet of the soil is where most of the moisture is absorbed.

of an inch)." Basically, roots can adapt to just about any environment.

Need more info? More proof? Take a look at Figure #57 [See page 94]. A young spruce tree in northern Finland maintains just over 64 percent of its feeding roots in the rotting duff, not even in the soil. What the tree wants, the roots go after.

But, trees still need "dirt." In his writings, Professor Thomas Perry makes it clear that trees on soils as little as five inches thick produce only poor tree and shrub growth, You can get fair growth with a ten-inch depth, good growth at 16 inches, and excellent growth with 20–30 inches of topsoil. (Remarkably, according to Perry, tree vigor is likely to gradually *decrease* with soil deeper than 30 inches.) One can then imagine

how widely and laterally the roots of trees growing in shallow soils must range in order to gather sufficient moisture and nutrients.

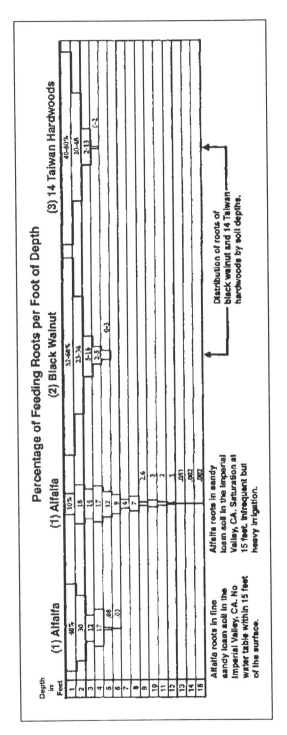

Percentage of Feeding Roots per Foot of Depth

| Depth in Feet | (1) Alfalfa | (1) Alfalfa | (2) Black Walnut | (3) 14 Taiwan Hardwoods |
|---|---|---|---|---|
| 1 | 40% | 10% | 52-68% | 40-60% |
| 2 | 30 | 18 | 23-34 | 20-48 |
| 3 | 12 | 14 | 5-16 | 2-13 |
| 4 | 17 | 17 | 2-5 | 0-2 |
| 5 | .08 | 12 | 0-3 | |
| 6 | .03 | 9 | | |
| 7 | | 6 | | |
| 8 | | 7 | | |
| 9 | | 2.6 | | |
| 10 | | 3 | | |
| 11 | | 2 | | |
| 12 | | 1 | | |
| 13 | | .051 | | |
| 14 | | .002 | | |
| 15 | | .002 | | |

Alfalfa roots in fine sandy loam soil in the Imperial Valley, CA. No water table within 15 feet of the surface.

Alfalfa roots in sandy loam soil in the Imperial Valley, CA. Saturation at 15 feet. Infrequent but heavy irrigation.

Distribution of roots of black walnut and 14 Taiwan hardwoods by soil depths.

**Figure #55:** Here are the nutrient levels of absorption for a black walnut tree, two alfalfa plants, and Taiwan hardwoods.

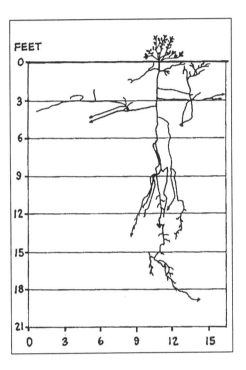

**Figure #56:** This mesquite has shallow roots to gather moisture from shallow soils after flash floods. The deep roots gather moisture that has had a chance to sink in after a gradual rainfall.

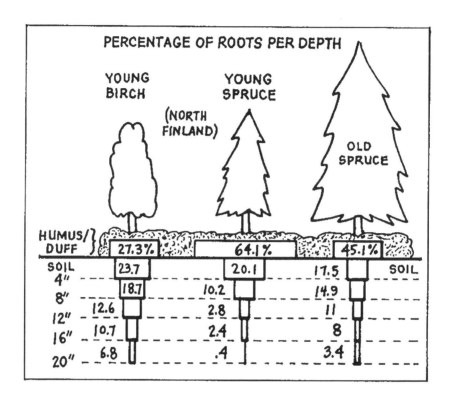

**Figure #57:** This amazing illustration shows how roots can grow up to feed in the duff in order to absorb nutrients that are as "fresh" as possible.

## ℭℛ PRACTICAL TIPS FOR GARDENERS

Here we go again: mulch, mulch, mulch. Replicate the duff that forms in a natural forest. Establish as many permanent pathways as possible. Try to let the pathways "breathe"—allow the air to flow into the roots and the carbon dioxide to be expelled. You can use chipped bark, chipped tree trimmings, gravel, or whatever local material suits you. There are paver blocks available in open cut-out hole patterns; these can carry nearly as much weight on top as a solid paver but look more attractive, as they add interesting texture, and grasses or ground covers can be planted to grow through the holes. Such pathways are great for wheelbarrows, working boots, and dress shoes with sensible heels (no spikes, please, unless the walker is willing to ruin a pair of high heels to help you aerate the soil).

FOOTNOTE

For scientific references, see: Zimmerman, M. H., and C. L. Brown. *Trees, Structure and Function*, New York: Springer-Verlag, 1971; Perry T. O., The Ecology of Tree Roots and the Practical Significance Thereof, *J. Arboriculture* 8 (1982):197-211; Sillick, J. M., and W. R. Jacobi *Healthy Roots and Healthy Trees,* Fort Collins, CO: The Cooperative Extension Resource Center, Colorado State University, 2003; Wray, Paul, and Amy Kuehl, *Tree Growth*, F-308/ Ames, IA: Iowa State University Extension. revised 1999; and Kopinga, J. *Research of Street Planting Practice in the Netherlands*. Proceedings of the Fifth Conference of the Metropolitan Tree Improvement Alliance. Pennsylvania State University, University Park, PA: 1985.

FOOTNOTE

# CHAPTER 10

# The Good Fungus Among Us

According to experts on the subject, the intriguing life-forms known as fungi comprise about 52% to 55% of a forest's biomass. Since a whopping 95% of all green plants depend on at least one fungal relationship in order to survive, it's appropriate here to discuss the magical relationship of beneficial fungi with the roots of trees as well as with a range of annual, herbaceous, and woody plants. This relationship is known as mycorrhiza, or fungus root, from the Greek: *mykes* [mushroom] and *rhiza* [root]. The plural is mycorrhizae.

Fungi, in general, form masses of tiny filaments known as mycelia, which frequently interact with plant roots. There are two major kinds of mycorrhizae: *ectomycorrhiza* and *endomycorrhiza*. In ectomycorrhiza (usually abbreviated as "EM"), these filaments remain outside of the plant, living on the cells of the root hairs. With endomycorrhiza, these filaments actually live between and inside of the cells of the feeding roots. There are numerous kinds of tongue-twisting endomycorrhizae: arbutoid, monotropoid, ericoid, orchidioid, and vesicular-arbuscular or arbuscular mycorrhiza, all of which interact with plants in different ways. The most important and widely distributed type of mycorrhiza is the vesicular-arbuscular mycorrhiza or arbuscular mycorrhiza (abbreviated as "VAM" or "AM".) (Some plants utilize neither EM nor VAM; examples of plants with no mycorrhizal association include all the species of brassicas—cabbage, broccoli, cauliflower, Brussels sprouts, etc.)

All mycorrhizal associations are beneficial and are characterized by the movement of plant-produced carbon to the fungi and fungal-acquired nutrients to the plant. Rather than a parasitic relationship, it is a mutualism in which both life-forms benefit. In general, most plants are dependent upon this union, as it is estimated that about 80% of all plant species in the world are mycorrhizal symbionts.

In general, mycorrhizal plants are well-fitted to endure environmental stress. Nutrient-poor or moisture-deficient soils show improved capacity for supporting plant growth and reproduction when mycorrhizal fungi are present. As if to return the favor, the plant allows the mycorrhizal fungi to extract sugars, starches, proteins, and lipids from its lateral roots. (We'll go into more detail a few paragraphs down.) Mycorrhizal fungi may also improve water absorption, increase drought resistance, and exude substances that reduce infections caused by some soil pathogens.

Phosphorus is the most common nutrient transferred via VAM association into the root system of a plant, especially if it is growing in soil that is low in this essential nutrient. All trees need phosphorus but are not always able to absorb soluble phosphorus efficiently; they are thus dependent upon the mycorrhizal relationship. The mycorrhizal fungi produce phosphatase enzymes that breakdown phosphorus compounds. (The absorption of micronutrients such as zinc and copper is also improved by mycorrhizal association.) The extensive mass of a fungal mycelium produces a huge surface area that allows the fungi to "mine" a much greater amount of soil and duff than the root hairs of the tree are capable of exploiting on their own.

As an example, the mycorrhizae can increase the absorbing surface area of pine seedlings

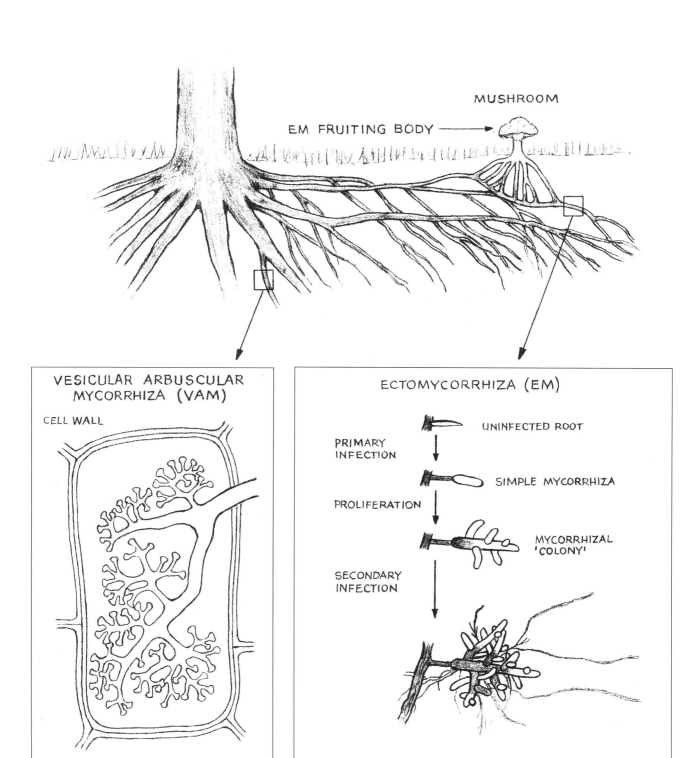

**Figure #58:** An illustration of both vesicular-arbuscular mycorrhizae (VAM) and ectomycorrhizae (EM) and how each differs from the other with respect to "colonizing" tree and shrub roots. The VAM fungi penetrates the root's cell walls, while the EM sends its mycelium in-between root cells. An "infection," albeit a beneficial one, then produces a colony of mycorrhizae to bring nutrients into the tree's roots.

by 80%. Some horticulturists maintain that the absorption surface area of a tree can be increased by 700-1000% by mycorrhizal fungi. Fortunately, there are plenty of these helpful organisms to go around. One scientist in Europe documented 101 species of mycorrhiza fungi associated with a single tree species—Norway spruce (*Picea abies*)—and 117 mycorrhizal species associated with Scotch Pine (*Pinus sylvestris*).

As mentioned above, the tree provides life-supporting sugars, starches, proteins, and lipids to the mycorrhizal fungi from its lateral roots. According to David Sylvia at Pennsylvania State University, "As much as 20% of the total carbon assimilated by plants may be transferred to the fungal partner." This transfer of carbon to the fungus has sometimes been considered a drain on the host, a plus for the mycorrhizal fungi and a small minus for the tree. However, the host plant may increase photosynthetic activity following mycorrhizal colonization, thereby compensating eventually for carbon "lost" to the soil. In the long term, a mutualistic mycorrhizal relationship is a winner for both the tree and the fungi. (A parasitic situation would exist if the tree received less benefit from the association than the fungi or vice-versa). Occasionally, plant growth suppression has been attributed to mycorrhizal colonization, but usually this occurs only under low-light or high-phosphorus conditions." In a study done with the sweet gum tree (*Liquidambar styraciflua*), Sylvia states the VAM fungi "might be used to alleviate" the problems of compacted soil.

Here's my favorite example. David Sylvia refers to M. H. Miller and coworkers from the University of Guelph, Canada, "[they] documented an interesting case in which disturbance of an arable, no-till soil resulted in reduced VAM development and subsequently less absorption of phosphorus by seedlings of maize in the field. Disturbance reduced both mycorrhizal colonization and phosphorus

absorption by maize and wheat roots but did not reduce phosphorus absorption by two nonmycorrhizal crops, spinach and canola. The authors concluded that under nutrient-limited conditions, the ability of seedlings with mycorrhizal association to release phosphorus may be very helpful for crop establishment."

As mentioned earlier, nitrogen is also transferred from the soil to the plant via the mycorrhizal association. Both VAM and EM help in the breakdown of nitrogen into a soluble form plants can absorb, especially in locations where plant litter is rich in lignin (the cells that make trees woody and sturdy) and tannins (astringent plant products). Only a few mycorrhizal fungi can mobilize nutrients from these primary sources. However, VAM and EM fungi can obtain nitrogen and other nutrients from sources of organic matter such as the biomass left in the soil from dead microbes and decayed roots, thus mycorrhizae may play an important role in nitrogen cycling in acidic and highly organic soils.

On the astonishing side, according to *Plant Roots: The Hidden Half,* by Yoav Waisel: "It is generally believed that…a large number of plant species are interconnected…by a relatively small number of fungal species." So, the forest really is all one living entity. This is the web of life at its best.

You can spot this wonderful interdependence with your own eyes; many ectomycorrhizal fungi produce fruit bodies near trees when the temperature and moisture conditions are suitable. Thousands of plant species have a relationship with EM fungi, as they tend to be specific to certain plants. VAM fungi are generalists, and the species with which they interact number only in the hundreds (perhaps as few as 150 separate fungi), but they form associations with more plants and trees since they are not specific to one host.

The fungi that form ectomycorrhizae are many common forest mushrooms. Well-known fungal genera are: *Amanita, Boletus, Cantharellus, Cortinarius, Hebeloma, Laccaria, Lactarius, Pisolithus, Ramaria, Rhizopogon, Russula, Scleroderma, Suillus,* and *Tricholoma.* Ask your local Mushroom Society to help you identify these mushrooms.

Here's an example as to how remarkably fungi can influence plant growth, from an article by David Sylvia in *Mycorrhiza Information Exchange*: "In 1955, soil from under pine stands in North Carolina was transported to Puerto Rico, where it was incorporated as inoculum into soil around one-year-old "scrawny" pine seedlings. Thirty-two seedlings were inoculated, and an equal number were monitored as noninoculated control plants. Within one year, inoculated plants had abundant mycorrhizal colonization and had achieved heights of up to 1.5 m, while most of the noninoculated plants had died. Further trials with mixtures of surface soil containing mycorrhizal fungi and with pure inocula, consisting of fungi growing in a peat-based medium, confirmed that inoculated seedlings were consistently more vigorous and larger than nonmycorrhizal ones. Subsequent surveys more than 15 years after inoculation indicated that the inoculated fungi continued to grow and [make spores] in the pine plantations."

## ∝ PRACTICAL TIPS FOR GARDENERS

Mycorrhizal inoculation with commercial products are not always necessary. We can assume that most garden soils have mycorrhizal fungi in them since most plant species are mycorrhizal hosts. However, the use of fertilizers, herbicides, and fungicides could diminish the mycorrhizal fungal populations.

If your property was bulldozed down to poor soil, if you're bringing in a lot of topsoil, if you've bought sterilized potting soil in a bag, if you live on a beach, if you have fertilizer or pH problems, soils low in organic matter, or are trying to plant in any highly disturbed site, then inoculating the soil with mycorrhizal fungi is important. Acidic soils low in phosphorus and zinc—and sometimes nitrogen, copper, potassium, and sulfur— often support good levels of mycorrhizae. However, well-fertilized soil does *not* support prolific mycorrhizal activity as the plants are getting enough fertility on their own.

Some garden suppliers now offer beneficial microbial and fungal inoculants to "boost" the soil's fertility, to be added routinely as one would any fertilizer. I called soil scientist Dr. Phillip J. Craul, Senior Lecturer at the Graduate School of Design at Harvard University, with the question, "What do the new packaged microbial and fungal products offer gardeners?" His immediate response: "I'm going to be frank; most of the time the home gardener doesn't need to fuss with [inoculants]. Such products are only helpful in special cases where you're working with sterile material; … most of the biological analyses show enough organisms are usually present…to repopulate the soil—even in urban areas." Furthermore, the inoculant you add probably wouldn't survive, as the soil's vast complex of biota tends to defend its own territory from "invading" microorganisms. According to Martin Alexander in *Introduction to Soil Microbiology*: "Microorganisms inoculated into non-sterile soil lead to poor growth, and often the seeded species is eliminated in a period of days or weeks [due to] a rivalry for limited nutrients; the release by one species of products is toxic to its neighbor, and [results in] direct feeding of one organism upon a second."

If you want to inoculate seedlings before planting them in containers with a sterile soil mix, you might try one of the inoculants on the market such as "Power Organics Mycorrhizal Root Booster" from Fungi Perfecti® LLC, at www.fungi.com/mycogrow/ or order "Mycorrhizal Root Inoculant" from www.harmonyfarm.com. [Available as of 2007 when this book was written.] A Google™ search will reveal many other suppliers. Try the key words: mycorrhizae inoculants, mycorrhizal fungi inoculants, and commercial mycorrhizal fungi.

FOOTNOTE

Trees such as alders *(Alnus* spp.), willows *(Salix* spp.), poplars *(Populus* spp.*),* and Eucalyptus *(Eucalyptus* spp. can have both VAM and EM associations on the same tree, while other trees, such as redwoods *(Sequoia* spp.) have a mostly VAM association.

FOOTNOTE

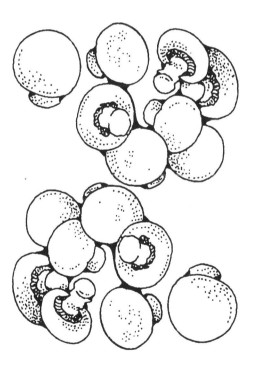

# CHAPTER 11

# Trees Water Their Neighbors (and themselves)

Here's a fact that amazed me when I first became aware of it: trees in summer-dry areas (or during a drought in normally moist climates) can "harvest" naturally occurring groundwater from soils six or more feet below the surface and "bank" the moisture in the drier surface soil for use during the following day. Lucky plants growing near the tree can also take advantage of this unusual source of moisture.

One of the first studies on this topic, using the sugar maple (*Acer saccharum*), was done in the early 1990s by Todd Dawson at Cornell University in upstate New York. The term "hydraulic lift" (HL), now the most commonly used designation for this moisture-banking phenomenon, was coined by horticulturists James Richards and Martyn Caldwell, during their work with Great Basin sagebrush in the late 1980s. They referred to the water that the sagebrush lifted as "hydraulically-lifted water" or "HLW." The process is thought to be a passive one, meaning that it occurs when the surface soil water content is lower than the water contained in the xylem of a root growing in that soil. [Xylem are the root's water-conducting cells.] Water obeys its natural tendency to flow from an area of higher concentration to one of lower concentration, in this case through the process of osmosis through the xylem walls to the soil. (Dawson's HLW study was conducted during a dry spell; once it rained, he noted, the HL cycling of moisture

stopped, since water was then readily available to roots growing in the surface soil.)

Additional study with sugar maples revealed that when HLW is combined with moisture provided by rainwater, understory plants (those growing beneath and around tree foliage) may utilize as much as twelve percent HLW. Thus, in areas where the soil's moisture level remains low, plants growing closest to the maple may have a natural competitive advantage over plants growing further from the tree. In fact, Dawson found that the closer-growing plants used from three to sixty percent of the HLW provided by the sugar maple.

In one study that concentrated on large trees, the amount of water "lifted" from the groundwater to the surface soils each night measured an amazing nine to eighteen gallons. It's also been noted that trees utilizing HLW demonstrate enhanced growth of both roots and leafy shoots. This may be due to the fact that the HLW facilitates the microbial process that helps the roots liberate nutrients from the soil, thus more nutrients are released and made available to roots growing in the upper levels. As of 2005, according to Dawson, over sixty species of shrubs and trees, in tropical, moist, and dry climates, have been documented as exhibiting HLW activity.

In a paper published in 2000, Stephen Burgess and his colleagues, using the Australian plant *Banksia prionotes* as an example, revealed that moisture can, as needed, also move downwards within the tree, laterally from moist surface soils to drier soils, or upward through the roots to surrounding drier soils. This system gave rise to their term "hydraulic redistribution," (HR) which includes the passive transport of soil moisture across different soil layers by plant roots. See Figure #59, based on an illustration by Todd Dawson, et al.

Even in moist Pacific Northwest forest, the

impact can be amazing. One study conducted simultaneously by J. Renee Brooks, et al., in a dry ponderosa pine (*Pinus ponderosa*) forest and in a moist Douglas-fir (*Pseudotsuga menziesii*) forest, revealed that, "A wide variety of understory plants use water redistributed from the overstory trees as far as 5m [16 feet] away from the tree." The study also found that, in the dry period of August, young (20-year-old) Douglas fir trees received up to 28% moisture, on a daily basis, from the upper 2m [6.5 feet] of soil solely from water redistribution at night; with old-growth ponderosa pines it was 35%. After 60 days of drought, hydraulic redistribution allows 21 days' worth of stored water in the upper soil horizons to be available to plants, near ponderosa pines, and 16 days near Douglas-fir stands, significantly reducing the drought impact on shallowly rooted understory species" although the effect was "patchy."

Interestingly enough, the effects of HR change over time. In July, the scientists noted, ponderosa pine tree roots that were six-and-one-half feet deep or deeper and contributed 60% to the daily flow of leaf evaporation. In September, however, the flow was down to 25% and the moisture was being lifted from the six-and-one-half-foot level to the soil four feet below the forest floor. Averaged over the summer, the water from the upper twentyfour inches of soil provided the daily movement of water through the canopy at about 12–14%.

In more current studies, Brooks et al., state: "Since 90% of the roots are in the upper 20 inches of soil, HR is extremely important for maintaining root function during droughts, and likely increasing surface root lifespan. This is particularly important for small understory plants and seedlings where 100% of their roots occur in these dry surface soils." In addition, these scientists found that the mycorrhiza benefit from HR and also contribute to transporting HR water to smaller plants. These surface roots and fungus are extremely important for nutrient uptake.

In the drier portion of the southeastern USA sandhills, researchers studied longleaf pine (*Pinus palustris*) and the turkey oak (*Quercus laevis*). One conclusion they came to "… suggests that tree roots can potentially contribute significantly to the rewetting of the topsoil."

A bit later in the research paper by J. F. Espeleta, et. al., they comment "Because tree and grass species are all mining the coarse sands of the topsoil, hydraulic-lifted water may significantly ameliorate the frequency and intensity of surface drought."

And, it's not just water that's moving around. Other researchers (R. L. McCulley, et al.) found, in the arid and semi-arid southwestern USA, that "…Hydraulic redistribution of shallow surface water to deep soil layers by roots may be the mechanism through which deep soil nutrients are mobilized and taken up by plants." Later they add "…[the] nightly partial recharge in the upper portion of the soil profile and the resulting delay in soil drying are consistent with a role of hydraulic redistribution in maintaining nutrient uptake by, and microbial activity around, fine roots."

And, you thought trees were just napping at night. Hah!

## ❧ PRACTICAL TIPS FOR GARDENERS

According to Brooks: "Any deep-rooted plant is capable of HR. Encouraging deep roots or interplanting deep-rooted crops can help supply small amounts of water to surface roots. This could be useful for keeping shallow-rooted plants viable during drought in times when irrigation is not possible."

It's good to know trees and shrubs do work together just as mycorrhizae helps the forest.

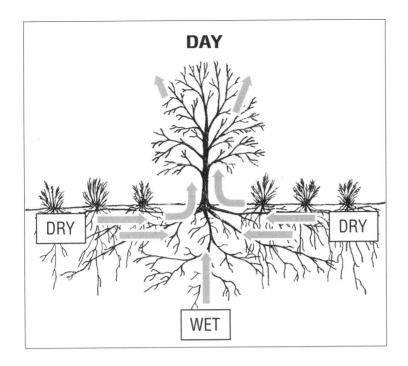

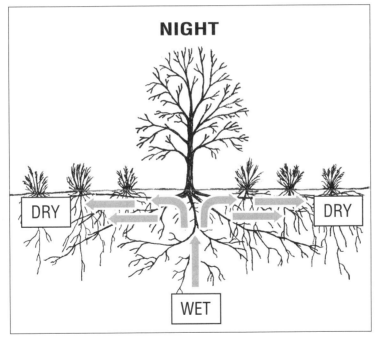

**Figure #59:** Trees can make moisture available to surrounding plants even as they store moisture at night in the upper soil levels for use the next day.

(Based on an illustration in: *Hydraulic Lift: Consequences of Water Efflux from the Roots of Plants.* Caldwell, Martin H, Todd E. Dawson, and K. James H. Richards. page 152.)

# CHAPTER 12

# Trees & Hardscape

[Given that the basics of root growth have been covered, the remaining chapters of the book are devoted primarily to gardening tips, suggestions, and ruminations.]

## Trees with shallow roots.

Here are some general guidelines to allow adequate space for root growth [there are *always* exceptions]:

- It's best not to plant trees between paved areas that are less than three feet apart.

- When there are only three to four feet between paved areas (such as between a sidewalk and the street), plant trees that mature at a height of less than thirty feet tall.

- If there is a space of five to six feet between paved areas, you can plant trees that will grow to as much as fifty feet at maturity*. [Recommendations for *minimum* planting widths between hardscapes or foundations, when noted with asterisks (*), have been provided by the city of Chico, CA, based on the performance of each tree as a street tree.]

- When planting trees higher than fifty feet, you'll need at least eight feet between paved areas.

You might also want to consider installing expansion joints in driveways, sidewalks, or patios near your trees. Expansion joints compensate for the contraction and expansion of the concrete in hot and cold temperatures; they are frequently used on public sidewalks, and are usually made of fiber, sponge rubber, plastic, or cork materials. They are typically dark-colored and look like dense roofing paper. They must be highly resilient and must not protrude above the concrete when contracting or become brittle in cold weather. They must also absorb the expansion of the two adjacent concrete slabs during hot weather. Look in the phone book for suppliers under the listing "Concrete Products," or check out places that sell ornamental rock and cement for tips and products. They can tell you how to use the material and how thick the concrete needs to be in your area for an adequate paved surface.

If roots should end up heaving parts of the hardscape, the use of expansion joints will limit the necessity for potential hardscape replacements to just a few sections instead of involving large areas. You might also be able to reduce heaving by shaping sidewalk sections near trees into a more narrow, curved pattern to accommodate root growth. Laying a bed of coarse gravel beneath pavements can be effective at slowing, or even stopping, the heaving action, because the tree roots will tend to grow downward into the earth, rather than up into the drying air pockets of the gravel. As mentioned earlier in the book, you can also make paths using lawn pavers with holes in the middle; these are easy to handle and will allow roots more "stretching room."

## Sidewalk-Friendly Trees

Here are some recommendations for street trees that require minimum planting-area width. [Again, recommendations from the city of Chico, CA, are marked with an asterisk (*); other lists of acceptable street trees are from St. Helena, CA; St. Louis, MO; the state of Iowa, and Seattle, WA.] These trees, while they are considered "sidewalk-friendly," can still cause problems if planted too close to paved surfaces such as patios, concrete or brick pathways, and foundations. The trees allowed in street-side plantings may be very different from area to area and in different growing zones; check with your local planning department for appropriate trees and regulations concerning street-side planting. Be certain of the mature height, spread, and cold- and heat-tolerance of trees before planting. Check to make sure the mature tree won't interfere with phone and power lines, or encroach on your neighbors' space. Ask your neighbors what hasn't worked and talk to local Cooperative Extension people.

## Trees for the Area Between the Street and the Sidewalk or Near Paved Areas (Hardscape) Taken from the various street tree approval lists mentioned above.

**Common name, *Latin name*, cultivar (if any) followed by comments, Zones (using the USDA hardiness zone map), and H x W.** [All trees are deciduous unless otherwise noted.]

**European Hornbeam**, *Carpinus betulus* 'Fastigiata'. Although the name 'Fastigiata' usually indicates a columnar shape, this tree matures to an oval-vase shape, with deep green leaves and smooth silver-gray bark. This sturdy tree with strong branch attachment prefers a seven-foot area between hardscapes.* Susceptible to scales. [Scales are insects that resemble stationary bumps on twigs, sometimes on leaves. They suck enough plant juices that the leaves may turn yellow and fall prematurely, and can cause twigs to die.] Zones 4–7(8), 50'x 40'.

**American hornbeam**, *Carpinus caroliniana*. When first planted, the shape of the hornbeam's canopy is pyramidal; as it grows older, it becomes more rounded. A good tree for smaller spaces, as it needs only four feet as a minimum planting area.* The bark has an interesting shaggy look. Sensitive to salt. Zones 4–9, 35'x 35'.

**Cornelian cherry**, *Cornus mas*. Oval, rounded canopy of dark-green leaves that turn a muted purple-red in the fall. One of the first trees or shrubs to bloom in the spring, as early as March, with bright-yellow blossoms. In mid-summer its red fruits are eaten by birds. No serious pests. Tolerates alkaline soil. Zones 4–8, 25'x 20'.

**Green ash red ash**, *Fraxinus pennyslvanica (F. lanceolata)*. Tolerates high pH, wet soils, dry soils, and hot, arid winds. Used as a windbreak and very commonly as a large street tree. Yellow fall color. Needs a minimum of seven feet of width for a healthy root system.* Do not plant both a male and female tree, as the resultant seed pods will be abundant and cause quite a bit of litter. Zones 3–9, 55'x 40'

.

**Ginkgo**, *Ginkgo biloba* 'Fastigiata' and 'Autumn Gold'. An ancient tree, literally a living fossil, with "roots" that go back some 150-200 million years. Each fall provides a glorious display of hot-yellow fan-shaped leaves. Be sure to buy a male tree such as 'Fastigiata' to avoid the female tree's smelly and messy fruit. 'Autumn Gold' is a male tree with spreading canopy. Ginkgos are pest- and disease-resistant, and are good city trees as they tolerate ozone and sulfur-dioxide pollutants. Ginkgos are resistant to oak root fungus but have only medium tolerance for salt used in winter de-icing. Prefers at least seven feet of width for its root zone.* Zones 4–9, 60' x 40'.

**Thornless common honey locust**, *Gleditsia triacanthos* var. *intermis* 'Shademaster'. A street tree with bright green foliage. Prefers seven feet of root width.* After the leaves fall, this tree's many long, twisted fruit pods can create a mess as they fall from the tree later on in the season. Good yellow fall color and medium sensitivity to salt from winter de-icing. When first planted, bark on new trees is tender and tends to get sunburned on the southwest side unless it's painted with white latex paint. Zones 4–9, 25'x 16'.

**Rocky Mountain juniper**, *Juniperus scopulorum*. 'Skyrocket'. Evergreen with blue-gray foliage. The species is pyramidal but rather wide, although 'Skyrocket' is narrower than most junipers. Can be used in a classic landscape design with parallel rows along a driveway or walkway to create an *allée*, or around a pool. Zone 3–7(9), 15'x 2'.

**Eastern red cedar**, *Juniperus virginiana*. Evergreen. Grows well under many conditions where other trees fade. Can be pruned to form a handsome hedge. Female trees produce berries that birds love to eat. Zones 2–9, 50'x 20'.

**Goldenrain tree**, *Koelreuteria paniculata* 'Fastigiata'. This wonderful tree blooms with bright-yellow flowers from June through late July. Also makes interesting pale-green seed pods that turn brown and look a bit frayed. Blooms three to four years after planting and makes a great tree for the patio or street—low to medium salt tolerance. Zones 5–9, 25' x 3".

**Crab apple**, *Malus* spp. Comes in many sizes, colors (with yellow-red fall foliage), and shapes. Some varieties are disease-resistant. Medium to high tolerance to salt. Birds are attracted to the fruit. Many crab apples need five feet of width for a planter area.* Zones 4–9. To learn about individual types and possible choices, consult with your local nursery.

**Persian parrotia**, *Parrotia persica*. One of the top ten trees for fall color, with its showy yellows, oranges, and scarlet reds. The new leaves even display nice shades of red-purple before turning to a glossy green, and trunk and limbs are covered in a fascinating brown shaggy bark. With a well-drained soil and a bit of moisture, this tree is virtually free of pests and diseases. Prefers slightly acidic soil, but will tolerate some alkalinity. Zones 4–8, 30'x 20'.

**Chinese pistache**, *Pistacia chinensis*. This is not the tree that produces pistachio nuts but in the same genus; the nut-producing species is *Pistacia vera*. The Chinese pistache is a popular ornamental street tree in the west and produces fantastic fall colors in a mix of hot, almost Day-glo™ oranges and reds. Be sure to buy a male tree if you don't want red berries to drop onto your patio. Takes drought well and is relatively free of pests, as well as resistant to oak-root fungus. If overwatered in a lawn area, it may develop verticillium wilt. Zones, 9–10, 40'x 40'.

**Willow oak**, *Quercus phellos*. Needs at least six feet between hardscapes to grow well.* Considered to have the most graceful foliage of all the oaks. Brown leaves may hang on the tree though the winter. Established trees are drought-resistant. Zones, 6–9, 50'x 35'.

**Scarlet oak**, *Quercus coccinea*. *Q. rubra* and *Q. palustris* are often confused with this tree, but the *coccinea* species has a more glorious red fall color (which may last up to one month) than either. Most large oaks need at least a 7' wide distance from hardscape.* Give each of these oaks plenty of room. Zones 4–8(9), 70'x 45'.

**Pin oak**, also known as swamp oak, *Quercus palustris*. A very popular street tree, with red-gold fall leaves, which in milder climates turn brown and hang onto the tree for most of the winter. One of the better oaks for growing in a lawn. Zones 4–7, 50'x 30'.

**English oak**, *Quercus robur* 'Fastigiata'. It's said by many that this one of the best-looking oaks if you want a more columnar tree. The species form is very grand and noble-looking—up to 60' tall and 35' wide. Most large oaks need at least a 7'-wide planting area.* Has high resistance to salt, but this cultivar and the species are prone to mildew. Cultivars 'Rose Hill' and 'Skyrocket' are more resistant to mildew. 'Fastigiata' is 60'x15'. All grow in Zones 3–7.

**American arborvitae**, *Thuja occidentalis*. Evergreen, and a common tree in home landscapes—because of its "well-behaved" roots, it is often planted near houses. This very sturdy tree can be sheared to form a low hedge or windbreak. Crisp, rich green in the spring and summer, fading to a brownish-green color in the winter. Hardy to –40 F. Takes all kinds of soils. Zones 2–7, 50'x15'.

**Little leaf linden**, *Tilia cordata*. Prefers a cool climate, can't take heat. Used as a common street tree in zones where it thrives. Is prone to pests and diseases, especially Japanese beetles, and aphids with their subsequent sooty mold. Zones 3–7, 60'x45'.

## Trees Incompatible With Paved Areas:

**Common name, Latin Name, Cultivar, followed by comments, Zones and Height x Width.** (This information is provided for placement in woodland plantings or in large lawns—as per the earlier discussion of roots and lawns.)

**Norway maple**, *Acer platanoides* 'Superform'. Tolerates dry summer soils. Yellow fall color. Considered to be an invasive tree in Virginia. Ask your local Cooperative Extension Master Gardener about how it performs in your climate. Zones 3–7, 50'x 40'.

**Red maple**, *Acer rubrum* 'Franksred'. Vigorous growing tree, very invasive. Intense shades of orange-to-red fall color. Prefers moist areas. Some sources rate this tree as satisfactory for planting in confined spaces, while others maintain it needs at least six feet of planter width. Zones 3–9, 45'x 35'.

**Silver maple**, *Acer saccharinum*, The fastest-growing of all the maples but easily damaged in rain- and ice-storms. Well-known for ability to buckle concrete. The roots are also invasive and clog drainpipes. Yellow to rich red in the fall. Zones 3–9, 60'x 40'.

**Sugar maple**, *Acer saccharum*. Fall color reveals brilliant shades of yellows, oranges, and reds. Its small greenish-yellow spring flowers are followed by medium- to dark-green leaves. Needs lots of room to grow. Zones 3–8, 70'x 45'.

**European beech**, *Fagus sylvatica*. A large tree with large roots, often seen planted in big lawns as a shade tree. An attractive, smooth, gray bark is revealed after the golden-brown leaves fall in autumn. Zones (4) 5–7, 55'x 40'.

**White ash**, *Fraxinus americana*. Grows best in deep, moist soils and needs plenty of room for its roots. Fall color changes from reddish-purple at the outer, upper portion of the canopy to almost yellow in lower and middle canopy areas. Zones 3–9, 60'x 60'.

**Tulip tree**, *Liriodendron tulipifera*. A handsome specimen tree with stature and good yellow fall color. It is susceptible to drought, and not a practical tree to plant near walkways, patios or house foundations. Zones 4–9, 80'x40'.

**Sweet gum** or **liquidambar**, *Liquidambar styraciflua*. Nice fall color, with actual color depending upon the cultivar. The cultivar 'Rotundiloba' is fruitless, thus avoiding the obnoxious prickly "sweet gum balls." Well-known for having invasive roots. Prefers moist soils. Zones 5–9, 70'x 45'.

**Southern magnolia**, *Magnolia grandiflora* and *Magnolia* 'Little Gem'. Evergreen. The Magnolia is a native tree with gorgeous waxy-white spp. flowers, the diameter of which of which may reach 15". Blooms in late spring and sometimes off and on during the summer. The species is enormous, as are its roots. ('Little Gem' is a compact tree, 20'x 8' with three- to four-inch flowers; it grows well in narrow places or near patios, but is shallow-rooted.) Both Zones 6–9, *M. grandiflora* 60'x 50'.

**London planetree**, *Platanus* x *acerifolia*. Bark is a fascinating mottled brown, tan and pale-green. Has been listed as an approved street tree in some communities. If planted in a narrow space between two paved areas, will often heave concrete walkways and curbs. Zones 4–8(9), 80'x 70'.

**American sycamore**, *Platanus occidentalis*. (*P.* x *hispanica*) Native throughout the Eastern and Midwestern parts of this country, where it evolved as a riparian tree (growing along rivers and streams). Needs at least eight feet between hardscape surfaces or away from a single hardscape like a patio. (It also drops many large leaves in the fall, a drag with a patio below.) Best planted in deep, moist soil or in a well-watered lawn with deep soil. Zones 4–9, 80'x 80'.

**Lombardy Poplar**, *Populus nigra* 'Italica'. Used to a riparian habitat, the roots of this tree grow near the surface, and it should be kept away from patios or other paved surfaces. The surfacing roots can create a problem, as they tend to be scalped by lawn mowers. Used in Europe to create a stately *allée* in which trees line both sides of a carriageway or driveway (with plenty of room for the roots), creating a spectacular approach to a residence. However, it suckers profusely. Zones 3–9, 80'x 12'.

**Weeping willow**, *Salix babylonica*. Not used as a street tree, as its limbs cascade down to the ground. Its roots probably won't crack pavement, but this tree grows more vigorously and looks more in its element when planted near a pond or other water feature. In a lawn, the shallow roots will surface, vulnerable to scalping by a lawn mower. Zones 6–8, 30'x 40'.

Visit old cemeteries or your nearest arboretum to see how large some of these trees will grow in your climate and zone.

# CHAPTER 13

# Trees That Belong Far Away From (or In) Lawns

Some trees just aren't very compatible with lawns. If you simply must include in your landscape one or more of the trees mentioned below, they belong in their own section, perhaps in a miniature woodland with compatible shrubs, ground covers, grasses and herbaceous plants, and in the further reaches of the yard, assuming that you have enough room.

This list can also be read as a list of trees and shrubs to avoid planting anywhere near a paved surface or foundation, as their aggressive roots may heave and break up rock paving, concrete, and bricks. These species mentioned here are generally suitable only for large yards with plenty of room so their roots can almost literally "run free." There are exceptions to these guidelines (mostly because trees can't read), and most any tree, including those listed below, can be found growing compatibly with a lawn somewhere. The goal here is to try to place trees properly for the highest healthy growth potential. Ask your neighbors about what did and did not work for them, and contact your local Cooperative Extension for fact sheets on this topic.

## Examples of Trees Best Kept Away From Lawns:

The **Silver maple** *(Acer saccharum* and *Acer* spp.

maples), shares its traits, especially the tendency to produce surface-oriented roots, with most other maples. This tree evolved on flood plains and "bottom land," and thus will tolerate periodic flooding. Its roots tend to grow up into the more aerobic upper levels of the soil and can be so invasive that the silver maple must be planted far from lawns, septic tanks, or foundations. Deciduous.

**American sweet gum** or **liquidambar** *(Liquidambar styraciflua)*, is famous for heaving pavement. It is also best kept away from your lawn, as the surface-growing roots are easily scalped during mowing. Likes moisture. Older trees produce hundreds of pounds of spiny fruiting balls each fall; this can be a nuisance over pavement and, if the tree is planted in a lawn, requires power-mowing with a bag to suck up the fallen fruiting bodies. The cultivar 'Cherokee' produces few or no seedpods. Deciduous.

London planetree, *Platanus* x *acerfolia*. There is some dispute about this tree's rooting pattern. In some settings the roots don't surface to cause problems, yet many cities have had to deal with sidewalks heaved by older and larger planetree roots. The difference may be in the watering strategy, and it may be that too-shallow irrigation leads to surfacing roots and sidewalk damage. This tree tolerates almost any soil. An English study revealed that 50% of the damage to buildings from the roots of species of the London planetree in shrinkable clay soils occurred within the first 18 feet from the trunk, and the roots often extended in a 50-foot radius. Since the London planetree originated in riparian (near flowing water) habitats, the roots will surface to suck up as much water and nutrients as you can throw at them. While this tree can be grown in a lawn with some success, it does drop plenty of fuzzy fruit. The dust from the fruit and underside of the leaves is irritating to some people. Deciduous.

**Lombardy poplar** *(Populus nigra* 'Italica'), is famous for lining estate borders or entrance drives, but *be* very *careful when considering planting this tree*, as its roots can sucker easily, especially if the soil is cultivated. The Lombardy poplar does provide a dramatic vertical line for a large property but must be kept well away from all hardscape and cultivated plants. Deciduous.

**Willows** *(Salix* spp.). All species of this genus evolved on floodplains or along riparian habitats, and tolerate periodic flooding. Willow roots naturally gather near the soil's surface to "grab" any water they can get. Deciduous.

The **common bald cypress** *(Taxodium distichulm)*, when planted in very wet soils or standing water, will produce the famous cypress "knees" (knobby root forms that protrude above the soil or water). The cypress is certainly inappropriate for a lawn because it requires much more water. This deciduous conifer, with its well-behaved columnar or pyramidal shape, can be planted in groves in moist areas where few other trees will thrive.

…and a shrub to avoid planting in your lawn:

**B**loodtwig dogwood *(Cornus sericea)* can be a large, floppy spreading shrub, but some pruning will give it a better form. The fall color is not remarkable; its beauty lies in the dramatic deciduous stem color during the winter, when the stems from the previous year's growth are truly blood-red.

Other moist-area trees and shrubs that tend to produce lawn-troubling surface roots or suckers include: **red maple** *(Acer rubrum)*; **Christmas berry** *(Ilex verticillata,* a shrub or small tree)*; **pin oak**; **swamp oak** *(Quercus palustris)*; **huckleberry (***Vaccinium pallidum,* a shrub); **sugar maple** *(Acer saccharum)*; **serviceberry** *(Amelanchier* spp., a shrub);

**Juneberry** *(Amelanchier arborea,* a shrub); **Eastern dogwood** *(Cornus florida)*; **American beech** *(Fagus grandifolia)*; **white ash** *(Fraxinus americana)*; and **common witch hazel** *(Hamamelis virginiana,* a shrub).

## Trees Among the Turf

**T**here's a real trick to choosing the proper tree to grow in a lawn or yard. Tree and lawn roots are often incompatible for four important reasons:

1. Turf grasses choke off the natural exchange of gases between the soil and the atmosphere; this can lead to a slight dwarfing or stunting of the tree.

2. The forest floor in its natural state is a deep litter of decomposing leaves, twigs, logs, and critters both microscopic and large; it provides a welcome place for roots to forage for nutrients. Taking away this beneficial natural mulch or duff and replacing it with lawn can harm the tree by leaving the roots exposed. Under these conditions, tree roots often appear to "surface," only to be scalped when the lawn is mowed; this is a source of potentially debilitating wounds for the tree and a bumpy ride for weekend warriors on their riding lawn mowers.

3. Some trees are simply programmed by their genes to produce many large surface roots. Many of these trees are native to the banks of streams or rivers, while others grow more inland and in moist areas. Attempts to control these root systems with fertilization techniques, irrigation, or root barriers usually yield poor results.

4. The fertility and irrigation required to keep a lawn happy are often more than a tree needs, and subsequent tree pest and disease problems follow.

Choosing the best tree to be surrounded by turf requires addressing the above concerns. Here's a detailed look at each problem or dilemma.

# 1: A Thatched Roof on the Lawn Soil

The turf selected for most American lawns is a far cry from the meadow or prairie grasses found in many states or counties. Natural grasslands or meadows tend to be composed mostly of "bunch" grasses. Bunch grasses, as their name suggests, don't form a consistent mat of shoots but instead emerge as clusters of root crowns in randomly spaced clumps. [See the discussion of buffalo grass earlier in the book, page 32.] Thus there is space between these bunches for dead, grass stems and leaves to decompose and gases to pass back and forth between the air and soil.

All roots and soil microbes expel small amounts of slightly toxic or harmful gases like tiny "farts" in the soil. Dying root hairs and soil microbes give off carbon dioxide as part of their "last gasps." Even this natural production of $CO_2$, however, if it accumulates in the soil, is detrimental to living young root hairs, which require certain levels of oxygen to prosper. As discussed earlier, a well-textured soil contains an enormous labyrinth of minute soil pore aggregates; these allow the debilitating gases to pass out of the earth and the revitalizing air to seep into the soil. The buildup of thatch (undecomposed grass leaves and roots) in a lawn retards this important exchange of gases. As with humans, fresh air is preferable to farts any day.

Farmers in England have used knowledge of this phenomenon to their advantage for hundreds of years. Long before there were dwarfing rootstocks for cherry trees, English orchardists seeded their cherry orchards with a grass cover (called a "sward" in the UK) to dwarf the trees slightly while encouraging them to bear earlier in their lives. Some grasses, such as ryegrass, actually produce a mild root excretion that acts to stunt the growth of nearby plants. [This is called *allelopathy* by scientists.] Ryegrass lawns can do the same thing to your trees. While the effect is mild, the ryegrass "works" around the clock.

# 2: Replicating the Forest's Natural Duff

As mentioned in earlier chapters, any tree, even one with a taproot, will still produce most of its feeding roots near the soil surface; in a natural forest, it will usually do so within one foot of the surface. The surface of a forest is defined as the top of the duff (undigested, undecomposed leaf and twig litter), *not* the top of the soil. Remember Figure #57 [page 94], showing studies in Finland that revealed that a young spruce tree (*Picea* spp.) grew an amazing 64.1% of its roots *above* the soil, feeding in and on the still-decomposing litter? The same spruce tree, even when mature, maintained 45.1% of its roots on top of the soil and laced through the forest duff.

By planting a lawn beneath a tree, you're removing the natural thick duff and replacing it with a thin layer of turf and thatch. Many trees have roots that are genetically programmed to grow into the duff; therefore, these roots naturally show up on top of the soil amidst the turf, ready to be mangled by lawn-mower blades.

The solution to the problem of emerging roots, as well as that of grasses retarding the growth of trees, lies not only in knowing what tree to plant, but in knowing how to care for it after planting. The best compromise for combining forest trees with lawn is a "skirt" of loose mulch extending as far from the tree trunk as is practical within the confines of your landscape or yard. In one study, grass growing right up to a tree reduced the growth of a "normal" tree to 58% of its normal height, while the growth of another tree with sod two feet away from its trunk was limited to 89% of its natural height. In order to have no effect on growth, the lawn-free area had to extend 20 feet away from a tree's trunk. Replacing sod with rough-and-tumble mulch will make your planting look more like a natural forest floor and permit the soil to breathe. The lack of thatch and of compaction produced by foot traffic and lawn mowing (since the "doughnut" of mulch will reduce the area that needs mowing) will encourage the soil's natural flora and fauna to keep the soil crumbly and well-aerated.

The depth of soil also shapes the root system of a tree, regardless of its genetic tendencies. If your yard has only six inches of good loamy soil on top of a heavy clay soil or a rock-hard caliche (calcium-based hardpan), then the tree's *entire* root system will be in the upper six inches. This will force more roots to extend above the soil into the path of the mower's blade.

In many climates, the wind will bring in weed seed to infiltrate your mulch. This can be a pain to deal with, but there are a handful of solutions: pull out by hand (a good choice if you stay on top of it and know a good chiropractor); dump more mulch on the pesky invader; place newspaper over the weeds and add more mulch; use a propane blowtorch [See Figure #4.] to kill the seedlings as they first appear (NOT a choice in areas with rain-free summers!)

## 3: Genetic Root Control

There are some trees whose roots will tolerate being near the soil surface or even thrive there. Many of these trees originated in riparian habitats. Examples not mentioned earlier in this chapter include: alders (*Alnus* spp.), tupelo (*Nyssa sylvatica*), and birches (*Betula* spp.). Along the bank of a stream, the roots of these trees are frequently exposed by bank erosion or may actually grow out into the water. Such trees are likely to produce noticeable surface roots when irrigation is too shallow, but even deeper irrigations won't completely override their genetic tendency toward surface roots. These are also some of the trees to avoid growing in a lawn.

Other trees with surface roots, but not necessarily from riparian habitats, are: Siberian elms (*Ulmus* spp.); Tree-of-Heaven (*Ailanthus altissima*—if you're not already stuck with it, ***do not*** *plant this scourge under any circumstances—anywhere—as it's extremely invasive!*); mulberry (*Morus* spp.); ashes (*Fraxinus* spp.); coastal redwood (*Sequoia sempervirens*); and sumac (*Rhus* spp.).

For more information about specific trees with surface roots that heave sidewalks and patios, see Chapter #12.

## 4: Water for Lawns and Trees, A Side Trip About Watering

While trees occupy far more foliage per square foot of lawn/garden space than grasses, they have a much more extensive (though not necessarily deeper) root system.

For this example, in Pasadena, CA for the month of July, the ET rate for grass is 7.1 inches per month. Compare the requirements of each tree in Figure #60 with that rate to get an idea of where the tree belongs in the water-use/ET spectrum. Ask your local Master Gardener for the monthly ET rates for your climate. Or, try nearby weather station, a local television station's meteorologist, local fire department, local Junior College, or nearby University. Most trees can survive quite well with irrigation or rainfall that totals less than the optimal ET for lawn grass in the local climate. You don't really have to worry about the specific numbers, just remember they are good relative index of; drought-hardiness the lower the number is below the average monthly ET, the less the plant needs water or supplemental irrigation.

## Some Minimum Irrigation Rates Based Upon the Evapotranspiration Rate for Trees in Pasadena, CA

These numbers, from the WUCOLS List (Water Use Classification of Landscape Species) produced by the State of California Cooperative Extension, allow for a reasonably good-looking landscape that doesn't appear drought-stressed. Each number represents the lowest average number of inches per month of irrigation as compared to the standard 7.1 inches based on cool-season turf. Up to a point, more water favors more growth.

Figure #60 shows that you can water most trees at much less than the ET rate that a lawn requires and they will still remain healthy. If you want to conserve water for your lawn by lowering the irrigation rate even more, the tree or shrub may grow more slowly and bear fewer flowers or less fruit. It is always possible to irrigate at a rate greater than the ET if the soil is *very* well drained; you may even get improved growth without disease and pest problems.

It quickly becomes apparent that the water needs of a gorgeous green lawn are often not compatible with that of many trees. Too much water on tree roots can cause crown rot of the upper root system and stimulate disease and pests. If you're planting a tree in the middle of a lawn, always choose a species with a water requirement similar to that of your turf.

## Fertility for Lawns and Trees

Many lawn grasses are so domesticated that they don't grow well or look their best without consistent fertilizing. Most species of trees grow in a wide range of fertility, often thriving without any supplemental nutrients. If properly chosen for your soil type and climate, no tree in your yard (except perhaps in the rare case of some fruit trees) should need added fertilizer or nutritious mulch, although you may want to use just enough woody mulch to conserve some

| Common Name | Latin Name | Low .7-2.1" | Med. 2.8-4.2" | High. 4.9-6.3" |
|---|---|---|---|---|
| Black alder | Alnus glutinosa | | | 5.6" |
| Common hackberry | Celtis occidentalis | | 3.5" | |
| Eastern redbud | Cercis canadensis | | 3.5" | |
| Gingko, maidenhair tree | Ginkgo biloba | | 3.5" | |
| Honey locust | Gleditsia triacanthos | 1.4" | | |

| | | | |
|---|---|---|---|
| Jacaranda | *Jacaranda mimosifolia* | | 3.5" |
| Goldenrain tree | *Koelreuteria paniculata* | | 3.5" |
| Crab apple | *Malus* 'Profusion' | | 3.5" |
| Oleander | *Nerium oleander* | 1.4" | |
| Ornamental/Chinese pistache | *Pistachia chinensis* | | 3.5" |
| Pin oak, swamp oak | *Quercus palustris* | | 3.5" |
| Red oak, Northern red oak | *Quercus rubra* | | 3.5" |
| Willow | *Salix* spp. | | 5.6" |
| Coastal redwood | *Sequoia sempervirens* | | 5.6" |
| Chinese scholar or Japanese pagoda tree | *Sophora japonica* 'Violacea' | | 3.5" |
| Yucca | *Yucca* spp. | 1.4" | |
| Sawleaf zelkova | *Zelkova serrata* | | 3.5" |

**Figure #60:** This chart (pages 117 and 118) illustrates the lower limit at which trees can be irrigated while still appearing in good condition. A sub-shrub like lavender (*Lavandula* spp.) can still look good with only 25% of the water as listed in the California WUCOLS List (Water Use Classification of Landscape Species).

moisture and keep the soil cooler during hot summers.

Under a moderate scheme of fertilization, most lawn grasses require a total yearly application of one to four pounds of actual nitrogen per 1000 square feet; many people, however, use far more fertilizer than they need to. Most landscape-contractor guidelines call for only one pound of actual nitrogen per 1000 square feet when planting trees. Only if a tree's leaves, especially the older ones, turn yellow-green (which often indicates nitrogen deficiency), should the rate of nitrogen application be doubled to two pounds per 1000 square feet.

When you feed your lawn, however, the tree roots just below the surface will certainly grab as many nutrients as they can. Thus, trees in or near lawns may grow tall and spindly due to absorbing more nitrogen than they require. This makes them prone to more diseases, pests, and damage from wind- or ice-storms.

Nonetheless, with turf that needs little irrigation or fertilizer and the right tree, you can certainly plant shade trees within a lawn. (Be sure to seed or sod the lawn with shade-tolerant varieties of grass.) The trick is to choose those trees that are somewhat genetically "programmed" to produce deeper lateral roots and not as many surface roots. Look at established yards in your neighborhood for some good candidates for your own lawn and landscape.

Here are some trees and shrubs to consider (as with life, there are trade-offs):

**Native buckeye** *(Aesculus pavia)* prefers some light shade but tolerates moisture. Prefers good drainage. Deciduous.

**Pecan** *(Carya illinoensis)* requires deep, well-drained soils with even moisture. It is difficult to transplant because of its true taproot nature. Deciduous.

**Column juniper** *(Juniperus chinensis* 'Columnaris') tolerates high levels of moisture in a well-drained soil, but make sure the roots don't get overly wet as a result of being planted too close to a sprinkler nozzle. Evergreen.

**Star magnolia** *(Magnolia stellata, Magnolia kobus* var. *stellata)*, like most magnolias, prefers an acidic soil high in organic matter. An even rate of moisture is important. Deciduous.

See the previous Chapter (#12) for a more comprehensive list, as trees that don't heave sidewalks often have deeper roots.

# CHAPTER 14

# Selecting Trees & Shrubs

## Selecting Shrubs and Trees at the Nursery

Local nurseries here in Sonoma County just love to see me drive up —"Here comes Kourik to buy more of our undersized plants!" they chortle. Meanwhile, I go laughing all the way to my garden, knowing that, within six to twelve months, these so-called "undersized" specimens will usually have grown to be bigger than those big "good deal" plants, often bigger than if I'd bought the next-largest (and more costly) container plant.

The process of choosing nursery stock starts with checking the roots, including the size and type of container or root-ball. With containers, ask the nursery staff if you can slip the root-and-soil mass slightly out of the container to check for root-bound plants. If the plant won't give with a very gentle tug, then you already know it's probably root-bound (also called pot-bound). Next, I assess the relative proportion of the top (or foliage) to the extent of the root system. I also look for a sturdy trunk and an undamaged graft (if it's a grafted plant), evaluate the branch pattern and canopy volume, and judge the color and look of the foliage. This set of criteria ensures that I'm getting plants that are able to handle transplanting, with healthy undamaged roots for vigorous growth and, where trees are concerned, the formation of sturdy anchors for wind resistance.

## Containers for All Seasons

Container plants provide the luxury of planting almost anytime during the growing season. The main advantage for the nursery in using containers is that they provide a more or less stable environment that allows plants to hang around for weeks on end; this gives "flexibility" to inventory management. While containers allow for the widest possible selection of plants, there are some drawbacks; unsold plants may grow root-bound and unhealthy, thus, the customer must be doubly careful to avoid getting stuck with root-bound remainders.

Healthy, properly grown container plants, however, can be easily transplanted with little trauma. One reason for this is that in a container the plant and surrounding soil have grown together and can be moved as a unit, so that the trauma of adapting to a new environment is cushioned by this integrity of the root mass. Field-grown plants, on the other hand, usually suffer during transplanting, with many roots disturbed and severed.

Another benefit of container plants is that all the young root hairs that are responsible for the plant's uptake of water and nutrients, are held within the container, not left behind in the field as frequently happens with balled-in-burlap [also called b-in-b, or B&B—I'll use b-in-b] and bare-root stock. With container plants, the root hairs are in place, immediately ready to absorb water and nutrients so as to prevent transplant shock, while field-grown stock must re-grow these vital root segments. Some containers [see below] actually make transplanting even safer and less traumatic, by stimulating the growth of many fibrous root hairs for quicker water and nutrient uptake.

## The Importance of Fibrous Root-Tips

The more root-tips there are, the less shock there will be at transplanting, because the newly sited plant can immediately begin to absorb needed water and nutrients. The goal of a good nurserykeeper is to produce as much root-tip growth as possible, in order to ease transplant shock and maximize the growth of the plant in its new location. Research done with Japanese black pine seedlings by Robert Hathaway and Carl Whitcomb at Oklahoma State University showed that, "Seedlings grown in container… were larger after three months than two-year-old bed-grown (field-grown) plants, and continued to outgrow the bed-grown seedlings after one full year following transplanting." Whitcomb surmised, after further studies, that the response may have been due to the container's influence in stimulating maximum root branching. Whenever a root hits the bottom of a container, it will either stop growing or circle around the can's bottom. If the root-tip stops growing, or if its growth is delayed, more side-shoots form—thus giving rise to more root hairs.

## The Basic Nursery "Can"

Across the country, containers come in more shapes and sizes than Baskin-Robbins™ has flavors. The most commonly used style, at least out west, is the tapered plastic "can," shaped for easy removal of the plant and for easy stacking and re-use. These "cans" are size-rated according to their approximate liquid capacity, and range from one to three to five to fifteen "gallons."

The roots of a plant left in a pot too long are inclined to grow out to the sides of the can, down to the bottom, and then begin to circle the pot as they continue growing. Plants which have been kept in a container long enough to develop circling roots are likely to grow more poorly after transplanting, because the roots usually continue to circle the planting hole. At a retail nursery, always ask for permission to pull a plant from its container to check its roots; any plant that has circling roots at the bottom of the root-ball should be avoided like the plague.

One helpful anti-circling design modification is the introduction of cans with vertically ribbed walls. The ribs, which protrude inside the can, help to guide the roots down to the bottom without any circling in the upper zones of the container. However, if any plant is left in its can for too long, even in a ribbed container, its roots will begin to circle.

Some progressive nurseries are now offering four-inch pots. These have the great advantage of being far less expensive than one-gallon cans, but you still have to watch out for root-bound plants. With the smaller size, it's much easier to slip out the plant to inspect the roots.

## Balled-in-Burlap, for the Sake of Tradition

For much of the country, b-in-b stock has been the norm for shrubs and trees. When a field-grown tree is dug up for transplanting, a lump ("root") of soil is wrapped in a binding of burlap that serves as a temporary container to hold the soil together. (When large trees are involved, a wire basket may also be used to contain the burlap and soil.) B-in-b plants are dug in the early spring near the end of the dormant season. This allows the plant to be held in the retail nursery even after it leafs out and through the summer. (Plants sold in a dormant state, without soil around the roots, are called bare-root stock and cost much less because the labor of wrapping the ball of soil has been avoided.)

With b-in-b, the digging procedure removes up to 98% of the tree's or shrub's roots, a considerable setback for any plant. The ball of soil needs to remain more or less intact to protect what few roots remain. In order to ensure an intact ball, the stock is grown in very heavy soils, so the soil clings to the root mass. This means that b-in-b is always considerably heavier than container stock of the same size, which is an important factor when it comes to ease of handling and preventing lower-back pain.

I've never purchased a b-in-b plant, since container plants are the norm in the nurseries where I shop. To learn more about the pluses and minuses of b-in-b stock, I turned to professional-landscaper friends on the East Coast. Earl Barnhart and Helga Maingay work in the Cape Cod area and were formerly associates of the New Alchemy Institute. They, like I, much prefer container plants to b-in-b stock. "Container plants," according to Earl, "are more stable; we can leave them sitting around much longer than b-in-b plants. They give us much more flexibility between the time we buy stock and the time we plant it." For the home gardener, this translates into the convenience of buying unusual or special stock when it's available and being able to hold it until there's time to plant. Earl says he prefers b-in-b "for trees more than six feet tall and all evergreen plants and shrubs."

When I visited horticultural writer Lewis Hill some 10 years ago; proprietor of Hillcrest Nursery in Greensboro, VT, notes that the availability of b-in-b stock is fading: "Most of our stock is [now sold] in containers, the exceptions being evergreens such as conifers, spruces, arborvitae, cypress, and pines."

## Bare-root Plants, for the More Organized Gardener

"Bare-root stock" simply means plants with naked roots. Since a bare-root plant has no protective soil around its roots, it must be sold and planted before its first leaves begin to show. In California, bare-root season begins in mid December and runs through early March and is a time of great savings for the organized gardener. Throughout the country, bare-root trees show up in the local nurseries only during the months that immediately precede spring bloom and leafing-out. If you've planned your garden or yard in advance of winter and know what you want, you can purchase bare-root trees and save a lot of money compared to container or b-in-b stock.

Bare-root stock is dug out while fully dormant from mass growing-fields in the late winter. The process is, depending upon your disposition, either amazing or frightening to watch. An enormous strange contraption that looks like a cross between a tank, a tractor, and a science-fiction moon-walking machine straddles the row of trees in the wholesale growing grounds, cuts the roots below the ground, and lifts the tree and an attached section of root system from the earth. In order to allow the machine to clear the treetops in the rows, another machine is used to lop off the trees at four to five feet above the ground prior to the passage of the digging monster. On a visit to Dave Wilson Nursery in Hickman, CA, Robert Woolley, estimated that, when a digging machine is used: "With pecans, one half of the roots, and with other stock, perhaps 15 percent of the tree's roots are left in the ground." Woolley estimates that "In order to maintain a high quality of stock, and to compensate for trees that are damaged in the digging process, we grow 10 percent more trees than we plan to sell."

Once the trees have been dug out, they are "heeled-in;" that is, the trees are tagged for identification and lined up very close together in long rows by named variety, then their roots are covered with moist soil, sand, or sawdust. The moist medium keeps the roots from dehydrating. (Some trees are stored in large warehouses without any moist medium but are sprayed with water frequently enough to maintain moist, healthy roots.) Once delivered to a retail nursery, the bareroot trees are again heeled-in so as to keep the roots moist. All of this is critical because, even though the tree is dormant, its roots can dry out enough that it will either grow poorly or die altogether.

Bare-root trees are cheaper than b-in-b because the expense of balling the roots is bypassed. The weight of the ball is also eliminated in transport and mail-order shipping, and the nursery only has to hold the trees in stock for a few months. Bare-root stock is also cheaper than container stock because the expense of potting up the plant is avoided. Bare-root is the only way mail-order companies can afford to ship most dormant decidous trees and shrubs.

The reduced cost of this form of stock to the customer is an obvious advantage, but the buyer should beware of bare-root stock that is potted up into containers at the end of the ideal bare-root season, i.e., after the weather is warm enough to cause budding and leafing to begin. I've found that if bare-root stock is forced into a container and held for any length of time, the tree, especially in the case of fruit trees, just doesn't seem to respond well to transplanting. If I were to purchase a containerized bare-root fruit tree after mid summer, plant it, and then wait and plant a new bare-root tree the following spring, the new bare-root tree would, after no more than two to three years of growth, be as big, or bigger, than the older, containerized bare-root tree.

The ethos of "small is beautiful" has long been my guideline for selecting container plants, yet I found no scientific literature to back up my years of experience. For help on this, I turned to Richard Harris, professor of Environmental Horticulture at the University of California at Davis and coauthor of *Arboriculture: Integrated*

*Management of Landscape Trees, Shrubs and Vines.* I asked Harris for his observations on smaller versus larger stock. "The general observation," replied Harris, "is that smaller-size plants often outgrow larger plants, especially if the larger plant has been transplanted into several increasingly larger containers. In field trials in the late 70s here at Davis, we found that several species of trees grown from seed planted in an unirrigated field were larger after three years than trees seeded into flats, transplanted into liners [these are like tube containers] and potted up into one- and then three-gallon containers, [even if they] were well-watered. I would say that the most important concept is that, except in the event of vandalism or animal damage, the smaller the plant when transplanted into the landscape, the better will be its relationship to the environment."

So, how to choose the right-sized plant? An important guideline is the plant's growth habit. When I buy lavenders (*Lavandula* spp.) in one-gallon containers, the top's height above the soil mix is often close to half the height of the container, or even smaller. With an herb like chives (*Allium schoenoprasum*), the tops may be taller than the one-gallon container, but I choose plants that don't have too many shoot divisions— this indicates fairly recent introduction to the container. While many people buy trees in fifteen-gallon cans with eight- to twelve-foot tops, I opt for specimens six feet tall or smaller. Since top-growth characteristics vary so much from variety to variety, the best buyer's strategy, until you become very familiar with individual species' growth rates, is to make sure the plant was transplanted into the container fairly recently. Harris recommends that the buyer "check to make sure the tree appears healthy and vigorous and that it can stand without the support of a stake...."

New, healthy white, pink, or tan-colored root growth and a root mass that hasn't completely filled the soil-mix volume also indicate a recently potted, non-root-bound container plant. Dark-brown or black root tips may be dead, and the presence of a lot of them may mean the tree may not transplant well. When in doubt, go for a non-root-bound plant in a smaller container.

## Balled-in-Burlap Trees

If you decide on b-in-b deciduous trees, early spring planting is the preferred and optimal timing. When buying, make sure none of a plant's buds have swollen or opened; the initiation of bud break also produces a hormone that initiates new root growth at the cut ends of large roots in the ball. You want the buds to break after the tree has been planted, so that the new root hairs come into immediate contact with moist soil.

## Top-to-Root Ratios: Think Small

With the soil-ball as its only protection, the health of a b-in-b tree or shrub is dependent on the amount of soil around its roots while it's held in the nursery. Wondering how a smart shopper could judge if the ball was large enough to sustain the tree, I again consulted one of my favorite books, *Arboriculture: Integrated Management of Landscape Trees, Shrubs and Vines*. A chart provided by the American National Standards Institute describes nursery standards for the ratio of trunk diameter (caliper) to the root-ball's diameter. [See Figure #61. Figure #62 gives the healthy ratio of the trunk's diameter to its height.]

Good shoppers take pride in getting their money's worth. When it comes to purchasing healthy, high-quality nursery stock, however, many people are led astray by first impressions. All too often I've seen retail nursery customers pounce on a plant with cries of "It's so big! What a great deal!" They rave over the size of trunk, branches, foliage, flowers, while ignoring the

source of the plant's sustenance and its literal foundation—the roots. Sadly enough, there are plenty of methods retail nurseries can and do use to keep plants looking good in spite of damaged, weakened, or defective root systems. The picture changes rapidly, however, when these plants are brought home to your garden. Many plants with crippled or misshapened root systems either fail to survive transplanting, or, having barely weathered severe transplant shock, limp along for years as miserable examples of their potential. Many of these poorly growing specimens are then allowed to remain in the landscape or garden because it's human nature to hold stubbornly to the opinion that, "I paid good money for that plant, and I'm not about to tear it out!"

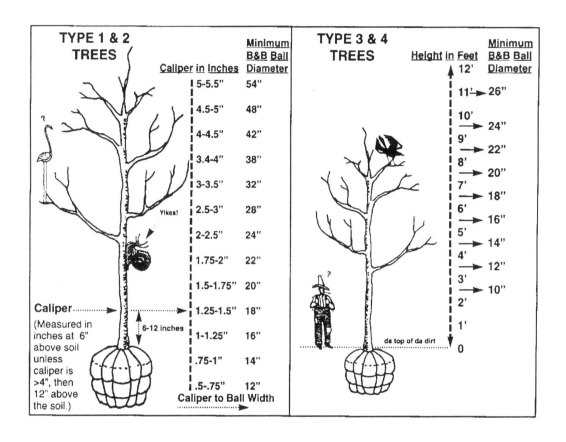

**Figure #61:** Use this chart to make sure you're not buying a b-in-b, tree (sometimes called B&B) that has overgrown its balled roots or has a trunk which is too small relative to its height. Fast- and slow-growing shade trees are covered by the chart on the left. Measure the diameter of the trunk (the caliper) six to twelve inches above the ball's surface. Find the caliper on the chart and read to the right to find the minimum diameter of the ball—bigger is better. Use the right-hand chart for small upright and small spreading trees. Measure the height of your prospective purchase. The chart shows the minimum size of the ball for survival and growth. For example, a 4.5' tall tree should have at least a 14" diameter ball. The bigger ball has more roots and may transplant with less shock and better growth.

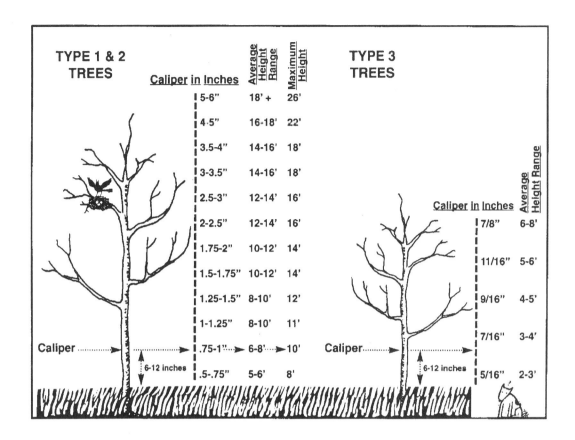

**Figure #62:** Whether b-in-b or container-grown, here is the ratio of the diameter of various types of trees to their average or maximum height. Measure the caliper at six inches (unless the caliper is greater than four inches—then measure twelve inches above the soil). Read up the dotted line to find the tree's caliper, then read over to find the recommended average and maximum heights.

# The Four Categories of Trees Illustrated in Figures #61 and #62

## Type 1–Shade Trees

| Botanical Name | Common Name |
| --- | --- |
| *Acer rubrum* | Red Maple |
| *Acer saccharinum* | Sugar Maple |
| *Betula* spp. | Birch |
| *Fraxinus americana* | White Ash |
| *F. pennsylvanica* | Green/Red Ash |
| *Gingko biloba* | Maidenhair Tree |
| *Gleditsia triancanthos* | Honey Locust |
| *Liriodendron tulipifera* | Tulip Tree |
| *Platanus* spp. | Sycamore |
| *Populus* spp. | Poplar, Cottonwood |
| *Quercus macrocarpa* | Bur Oak |
| *Quercus palustris* | Pin Oak |
| *Salix* spp. | Willow |
| *Tilia americana* | Linden |
| *Ulmus americana* | American Elm |

## Type 2 –Shade Trees, Slower Growing

| | |
| --- | --- |
| *Aesculus* spp. | Horse Chestnut |
| *Celtis* spp. | Hackberry |
| *Cladrastis lutea* | Yellow Wood |
| *Fagus sylvatica* | European Beech |
| *Koelreuteria* spp. | Goldenrain Tree |
| *Laburnum* spp. | Golden Chain Tree |
| *Nyssas sylvatica* | Tupelo |
| *Quercus alba* | White Oak |
| *Sorbus* spp. | Mountain Ash |
| *Tilia cordata* | Little Leaf Linden |
| *Tilia euchlora* | Crimean Linden |

## Type 3 –Small Upright Trees

| Botanical Name | Common Name |
| --- | --- |
| *Acer campestre* | Hedge Maple |
| *Acer circinatum* | Vine Maple |
| *Cercis* spp. | Redbud |
| *Halesia* spp. | Silver Bell |
| *Malus* spp. | most Crab Apples |
| *Prunus caerasifera* | 'Thundercloud' |
| *Prunus* spp. | Ornamental Plum |
| *Styrax* spp. | Snowbell |
| *Syringa amurensis* 'Japonica' | Lilac |

## Type 4 –Small Spreading Trees

| | |
| --- | --- |
| *Acer palmatum* | Japanese Maple |
| *Acer griseum* | Paperbark Maple |
| *Cornus* spp. | Dogwood |
| *Lagerstromia indica* | Crape Myrtle |
| *Magnolia soulangeana* | Saucer Magnolia |
| *Magnolia stellata* | Star Magnolia |
| *Malus sargentii* | Sargent Crab Apple |

# CHAPTER 15

# Planting Trees & Shrubs

The upper horizons of the soil are so critical to plant growth. Remember Figure #52 [page 90] in the chapter "Native & Ornamental Trees," which I described as one of the most important illustrations in this book? This drawing illustrates clearly the importance of even the top two inches of a forest's floor, and why the goal when planting a tree or shrub should be to preserve the integrity of the upper zones of the soil as much as possible.

I like to plant trees and shrubs on constructed soil mounds as opposed to planting them in flat ground. If you're working with slightly heavy clay/loam soils, the mounding is especially critical to preventing root or crown rot (*Phytophthora* spp.—a fungal disease of the upper portion of the roots, near the soil surface). Many ornamental and fruiting tree and shrubs die of crown rot without the gardener ever suspecting the culprit. The fungus damages the sapwood, either killing individual limbs or the entire plant. Once the symptoms (pale-yellow, wilted leaves) appear, it's too late to do anything about it. One clear indicator of *Phytophthora* in soil is the health of rosemary plants growing in it: if you see one or more rosemary "limbs" turn yellow, then you're dealing with root rot. There is no chemical way to kill the pathogen, and there are only three effective ways to deal with it: (1) remove dead and dying plants and replant with plants resistant to *Phytophthora* spp., (2) avoid overwatering, and (3) plant on mounds. (For a list of both resistant and susceptible plants, see Appendix #6.) Observant horticulturists have noticed that the trees that survived this disease were either planted in well-drained soil or established on mounds to protect the upper part of the root system. The mounds act as a preventative measure, especially in climates where it rains during the summer, because the crown-rot fungus thrives in damp *and* warm soil.

So many people are suspicious of the mounding method, (even though I've used it successfully with ornamentals and edibles for nearly 20 years) that I decided to call on my two favorite authorities on trees: Carl Whitcomb, formerly of the Oklahoma State University at Stillwater, who pioneered the research on reducing the use of soil and fertilizer amendments, and Alex Shigo, formerly with the USDA Forest Service, the man who taught the world how to prune properly and who certainly knows his roots. (This was before his untimely death on October 6, 2006.)

When I asked Whitcomb if he'd had any experience with planting on mounds, he replied: "Yes! At Oklahoma State University [Stillwater], we planted out some sycamores seven to eight years ago right on the ground, with no planting hole or depression. We just poured a large cone of topsoil over the roots, to create a steep angle. They needed more water the first several months during dry spells, but they rooted out well and one of them remains [the others were removed purposely] and has a five-inch caliper. It hasn't received any supplemental watering for seven to eight years."

Shigo's reply was: "I'm all for it. The best botanical gardens in England and Australia plant *all* their trees on small mounds. It's easier to regulate watering from a position of dryness than from wetness." When I asked him why the mounding technique hasn't caught on in America, he answered: "Our pioneering spirit still prevails.

It's easier to treat trees as expendable than to show them the respect they deserve. Now that we've run out of land and we're trying to plant in tougher settings like prairies [where trees don't naturally occur in great abundance], we're finding out how important techniques like this one are."

## No Amendments Is Good Amendments

Amendments—materials such as sand, peat moss, compost, and rice hulls—are often added to planting holes in a well-meaning attempt to improve drainage and keep the soil loose and friable. Fertilizers, such as blood meal, cottonseed meal, greensand, and wood ashes are frequently also added, out of a desire to provide nutrients. Some amendments, such as compost, are thought to both increase drainage and act as mild fertilizers.

One of the best studies on the effect of adding amendments to planting sites for fertility was done at Oklahoma State University at Stillwater by Joseph Schulte and Carl Whitcomb. They planted 108 silver maple trees (*Acer saccharum*), using 11 different soil treatments and a control (an untreated planting hole). One of their conclusions was: "No benefit was derived from the use of soil amendments either with a good clay loam soil or a very poor silt loam subsoil." They found that the control plantings with no additional amendments generally outperformed plantings to which drainage and fertility amendments had been added.

Under hard rains or heavy irrigation, the loose soil of the amendments in a traditional planting hole turn into something like an underground swimming pool, drowning the tiny root hairs that are so important for absorbing nutrients. Adding a lot of amendments to a planting hole also leaves the roots unprepared for the shock

of the unamended soil that lies beyond the hole. (And, nobody can amend the area of an entire mature root system, since it will extend much further underground than the dripline.) Often the roots fail to make it out of the well-amended hole and wind up circling around in the loose planting medium, making the trees extremely likely to blow over during a storm. Remember, the trees most tolerant to wind are those with the widest root systems.

In sandy soil, planting holes can tolerate the addition of more amendments because the contrast with the surrounding soil before and after amendment isn't so different. But, since the fiber of the amendments will absorb plenty of water, adding a lot of them to drought-prone sandy soils will, again, concentrate the roots in the more moist, amended area, and leave the tree vulnerable to wind damage.

## No Fertilizer Is Good Fertilization

Roots are relatively lazy; they feed where it's easiest. Fertilizers encourage the tree's roots to stay in the planting hole. Compost, especially in large amounts, turns out to be one of the worst additions when planting trees because it acts as both a "sponge" and a fertilizer. If you feel you must fertilize, add the materials as a top-dressing *beyond* the planting hole to encourage the roots to spread into the native, unamended soils.

## Step-by-Step Mound Planting

It should be noted that this technique assumes the gardener has chosen the right rootstock for the soil. For example, plum trees can handle some clay soils, peach trees can't.

When planting a bare-root tree or shrub on a 6-,

12-, or 24-inch-high planting mound, an actual planting mound isn't even required. (With some shrubs the mound can be shorter, proportional to the size of the plant.) Draw up or figure out planting formations in advance so you can get as many of your desired plants as possible in the form of bare-root deciduous tree stock. This keeps the price down and ensures virtually no transplant shock. For early-spring planting, it's always better to plant bare-root trees rather than container or b-in-b plants, because, as mentioned before they're cheaper. The same techniques I'm giving you here may work when the tree is in full leaf, but there is a great risk of transplant shock and losing the tree or shrub. In warm winter areas, plant evergreens early in the spring or late in the fall to avoid the unknown extremes of summer.

So, here's the ideal mound-planting technique: soak your bare-root tree in water while making the mound; this rehydrates the tissue and washes off debris clinging to the roots. Next, remove all the grass and weeds from the two- to four-foot diameter of the mound-to-be.

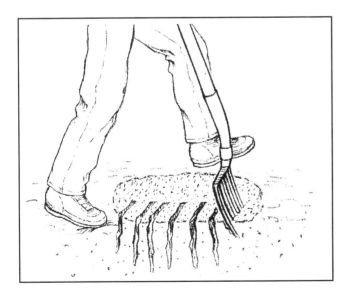

**Figure #63:** Heaving open the soil beneath the soon-to-be mound helps fracture the earth and make it easier for the roots to quickly penetrate the soil.

Use a spading fork to work the soil in order to keep from "slicking" the sides of the planting hole, and work with a motion that is more like just heaving, cracking, and breaking open the native soil within the entire circle. [See Figure #63.] Save your sweat and dig only as much as is necessary to fit the root system of your transplant. Remember, with this method, a planting mound is more important than the actual planting hole, which should only be as deep as those roots that won't be covered by the mound. Now, scrape up

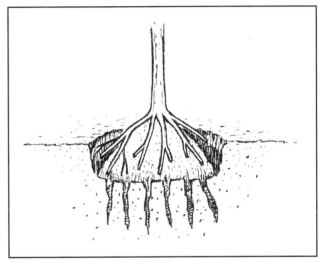

**Figure #64:** A small cone of soil is helpful when spreading the roots of a bare-root tree.

good soil from the surrounding area (after the weeds have been skimmed off) and make a small cone of soil in the center of the circle. [See Figure #64.] Then spread the roots of the bare-root tree over the top of this cone of soil.

Make sure some of the exposed roots are tucked into the native soil that was previously fractured open with the spading fork. After the roots of the trees are spread evenly out on the cone of soil, gather plenty of topsoil to make the rest of the mound. Make the final mound twice as wide as it is tall, as it will often shrink up to half its size as the soil settles. (Pile that dirt at least 12 to 24 inches high to achieve a final height of 6 to 12 inches.) Cover the bare-root tree's trunk to the same depth as it was in the growing grounds; this can be detected by noting a marked change in color on the trunk. Tamp the soil with your shoes to eliminate air pockets that can desiccate young roots. (The soil in the mound can be amended for better drainage so long as some of the tree's roots are placed in unamended soil.) Rake the finished mound to look like a gentle knoll.

In areas where the summers are dry, make your planting mounds in the fall; then all you have to do in early spring after the trees arrive is open the winter-moist soil enough to place the root and cover it with native soil. In areas where it rains in the summer, wait until the soil drains after a rain so that it's moist, not wet. Form a snake-like coil of soil with your hands; if it holds together, it's still too wet. If it crumbles damply, it's just moist enough.

FOOTNOTE

**Note**: If you live in an area populated by gophers, a bigger hole will be needed (as opposed to just fracturing the soil) to allow for inclusion of a protective hardware cloth basket. The bigger the basket, the more roots you'll protect from these hungry little devils. Be sure that the upper edge of the basket protrudes at least six inches above the soil and mulch to keep nocturnally wandering gophers from slipping inside your secured perimeter. To place a wire basket, simply dig a hole two to three feet deep and wide, put the basket in, and fill the hole again.

FOOTNOTE

After planting, soak the entire mound once or twice. Then mulch the mound and beyond about four inches deep, making sure that the mulch stays at least six to twelve inches *away* from the newly planted trunk. From that point on, water outside of the mound by placing drip-irrigation tubing in a loop around; this causes the roots to "explore" the surrounding native soil. A few days or weeks after transplanting, depending on the weather, arrange the loop to circle the base of the mound even more widely. Figure #65 summarizes the basics of a properly planted tree or shrub.

## Planting from Tube Containers

Trees and shrubs grown in variously shaped tubes or deep containers are especially useful when planting hedges, windbreaks, trees intended to last a long time, or large woodland-like plantings, and the cost is often lower than that of other containerized plants.

Tube-grown seedlings should come with a few air-pruned roots and plenty of lateral roots. A mound may not be required if the soil is well drained or matched to the rootstock. Mounding won't hurt, but takes extra time.

To plant, turn the tube upside down and jostle or tug the seedling out. Shake off as much of the potting soil as possible. Spread or fan out the roots. With a spading fork, heave open a crack in the soil and place the seedling's trunk in the ground at the same level it was in the tube. To close the planting hole, lift the fork out, place it 4–6 inches away from the seedling and parallel to the previous insertion, and press the handle toward the seedling to compress the soil around the roots. Water thoroughly and add a ring of mulch, leaving the trunk uncovered.

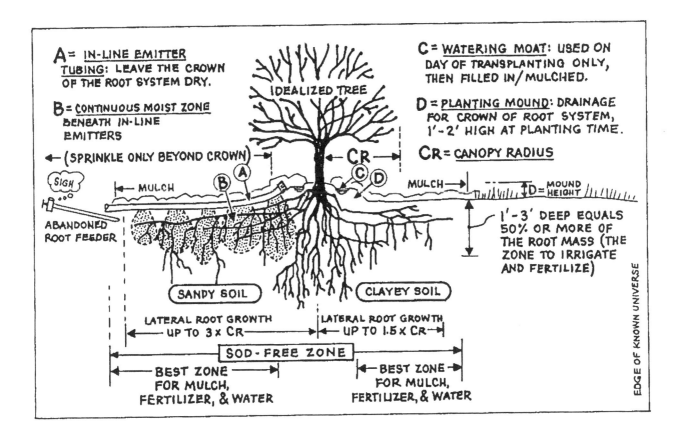

**Figure #65:** This is a detailed illustration of how one might plant a shrub or tree on a mound. It also illustrates how to irrigate on the day of planting by using a moat of water. Shortly thereafter, the moat is filled in and drip irrigation at the dripline begins, to be followed by wider and wider lines or loops of in-line emitter tubing.

(From: *Drip Irrigation, For Every Landscape and All Climates.*)

## Sunburn

Painting new trees, especially bare-root trees, protects trees from sunburn, which can lead to dead tissue and the invasion of fungi. Use a white or very pale beige indoor or outdoor *latex* paint. (It *must* be latex, not oil-based.) Paint from the very bottom of the trunk (pull the soil back a little when painting the bottom to make sure the lower part of the trunk is covered with paint; replace the soil when the paint has dried), and continue up to the first lower branch. The paint will disappear in several years, by which time the bark will be strong enough to take the sunlight. Painting can benefit all new trees, as they are usually grown initially in heavily shaded areas (a situation which leads to tender bark), due to the close proximity of other plants in the nursery area.

# CHAPTER 16

# Planting Root-Bound Trees & Shrubs

Sad but true: the mass-plant-bargain-buy industry appeals to inexperienced gardeners' very understandable desire to get the biggest plant for their money. This is why your chances of coming home from a home-center nursery or other retail outlet with a pot-bound container plant are fairly high. Few people realize, however, that a smaller plant purchased with non-crowded roots may outgrow a bigger root-bound plant in as little as one season. [See Chapter #14, page 121, "Selecting Trees and Shrubs."]

If you do wind up with a big root-bound "bargain," your only choice is to rid the plant of excess roots. Begin by tearing apart the root mass. This will cause less damage if the root system has been well-watered, so soak the roots for up to an hour to make sure the entire root mass is saturated. Use your hands to gently (the operative word) separate and tear open the root mass, starting at the bottom and carefully working your way up. If the plant is severely root-bound, you may need to use a knife or pruning shears to cut several lines down the side of the root mass before you begin to tear it open. As you separate the roots, try to leave the pinkish-white ones, as they are still capable of actively absorbing nutrients and moisture, but you can discard any brittle brown ones, as they are no longer viable.

If the soil you're planting into is any heavier than the potting soil used in the sale pot, try to remove as much of the original potting soil as is possible without letting the root mass fall apart. You don't want to turn the plant into a totally bare-root plant, so leave some roots and soil bound together.

Next, spread the roots apart as well as you can. This is an attempt to get a pot-bound tree or shrub to grow some healthy laterals, rather than just producing roots that continue to circle underground once the tree has been planted (spreading the roots may or may not work, but is certainly worth trying). Construct a small planting mound as recommended in Chapter 15 [page 133], and distribute the spread-out roots radially and evenly over the mound. You may have to prune off some of the twisted circling roots as you do this.

Cover the roots with native soil. Press firmly with your hands or feet to exclude any air pockets. Water thoroughly. No vitamin $B_1$ (sometimes recommended in garden manuals) is required, as studies show it doesn't really help. If you can't resist, it doesn't hurt the roots and doesn't cost much.

Next, prune back the foliage. Try to prune off no more than the same percentage of roots that you remove. Some disagree with this approach, but I've found that it works for me. If your transplant comes with ideal roots that don't need pruning, then leaving all the apical (tip) buds of the foliage is the best way to go, as they help send hormones to the root-tips to encourage growth.

## Thinning the Top

When selecting trees, the ideal is to avoid those with canopies out of proportion to their trunks. However, the selection at many nurseries may (almost inevitably) be less than ideal in that respect. Thinning out a top-heavy tree is one way to alter the "sail" effect of an overlarge canopy on a weak stem. (The canopy acts like a ship's sail, catching the wind and making the tree more likely to be easily bent or blown over.)

Horticulturalist Richard Harris suggests that transplants that are allowed to grow with little or no pruning will grow better than those that are severely pruned, but he also mentions that up to one-fourth of the canopy can be removed without severely affecting the tree. As a matter of fact, if drought is anticipated, severe pruning of the treetop can actually be helpful in reducing transpiration. On poorly grown, spindly trees on which the top is disproportionately taller and bigger than the root-ball, you can thin the treetop selectively by up to 25 percent after planting. The

thinning will help reduce the "sail" effect and allow for a better proportion between roots and the top.

Ideally, planting and thinning of deciduous plants should take place during the dormant season, since pruning at this time will stimulate more shoots and foliage. If you want to thin the canopy without encouraging new growth, do it in mid- to late summer. In this case, be sure to monitor your soil-moisture level carefully, because the unpruned top will transpire plenty of water before it's time to thin.

If it's necessary to stake a new skinny-trunked transplant to keep it from flopping, the lower you tie the tree to the stake(s), the more the trunk will be able to flex and strengthen in the wind. To determine the best height for tying (a height

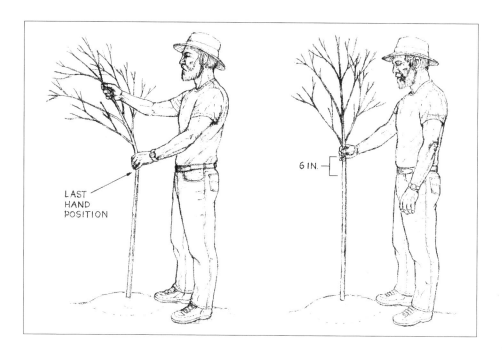

**Figure #66:** By simply bending a spindly tree, you can determine the best place to locate the stakes and tree ties. Allowing the tree to move as much as possible in the wind builds up more trunk girth. Trees that have been gently buffeted by the winds can be released from the stake. Trees tied so the wind can blow the canopy around can be untied earlier than ones tied up too high.

which will allow flexing but not flopping), grab the trunk with your right hand near the top and gently bend the top over with your left hand. You'll notice that the tree is able to return to an upright position. Move your right hand down the trunk and continue bending and releasing the top with your left hand. At a certain point, the top will stay flopped over and not regain its upright position. Tie the tree to the stakes at a point six inches higher than the last position of your right hand.

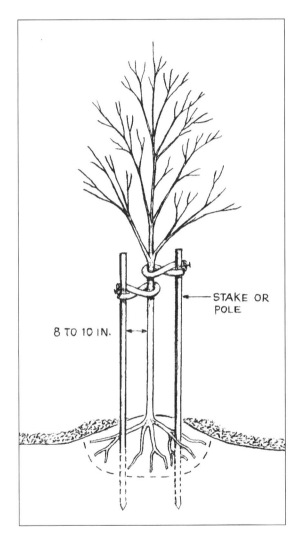

**Figure #67:** Use a tree tie from your local nursery to make the two figure-eight ties. Or, use a length of old hose with braided-wire sticking out both ends to twist together the two strands.

# Appendix #1

## Subsurface Drip Irrigation (SDI)—Putting it all Together

During geopolitical upheavals, going "underground" can be an important means of survival. Subsurface drip irrigation (SDI) is primed to take its place in the upcoming geopolitical issue of water rights and distribution and will no doubt have more and more influence on how we water our lawns, shrubs, trees, or plantings in narrow spots. The future is subterranean. Out of sight!

### Start With a Back-Flow Preventer

SDI can start at the faucet or any other plumbed connection to the water supply [See Figure #68.] The first gizmo is a brass vacuum breaker. Essential to protect the purity of your drinking water, it keeps dirt, manure, mulch or chemicals from siphoning back into your home's water system. Backflow preventers can be installed at each separate faucet, or a single one at a central location for all exterior water use.

In order to work, most backflow preventers (such as the inexpensive brass atmospheric vacuum breaker) must be installed at least 12 inches higher than all other parts of the drip system; this is no problem when the tubing is buried four to eight inches below a lawn which is

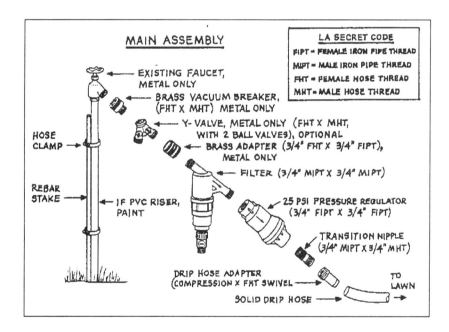

**Figure #68:** The main assembly must include an anti-siphon device, a filter and a pressure regulator. This basic main assembly is built onto an existing hose-bib which stands at least 12 inches above the ground-level tubing. This makes it easy to clean the filter during the irrigation season by using the ball valve on the filter cartridge.

level and below the faucet. A brass atmospheric vacuum breaker costs around $6 to $15, depending upon the model.

The safest anti-siphon device (and the only backflow preventer that can be installed below the level of the drip tubing) is a brass-bodied double check valve, a special backflow preventer that costs $150 to $200. *Be aware*: the double check valve can only be installed by a certified landscape irrigation contractor. Consult with the local building code department as to what model is approved for your area.

## The Filter

Without proper filtration, any emitter can clog. [Clogging is much less likely with the in-line emitters, which I'll cover a bit further on.] With most municipal water supplies, the filter needs only a 150-mesh metal screen, while well water requires a 200-mesh metal screen to handle fine sediment. (The higher the mesh number, the smaller the openings.) GEOFLOW™, a manufacturer of in-line emitter drip tubing, recommends a 300-mesh screen with their one-half-gallon-per-hour (gph) emitters and a 150-mesh screen with all one-gph emitters.

Y-filters with a ball valve are the best filters because the ball valve on the filter cartridge allows the gardener to flush sediment and sand from its screen with a twist of the discharge valve and without having to unscrew anything. Agrifim™ makes an excellent, sturdy Y-filter made of "filled polypropylene" (a soft non-brittle plastic) that withstands water pressure up to 150 pounds per square inch (psi).

## In-Line Emitters

In-line emitters are one of the least known drip-irrigation technologies, but offer (as I've discovered in my 25 years of gardening and landscaping experience) the best mix of efficiency, ease of installation, and resistance to clogging and leaking. The tubing is nearly

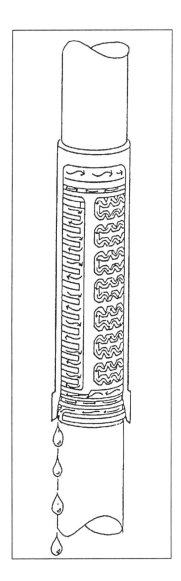

**Figure #69:** The beauty of an in-line emitter is that there is nothing to break off and it's clog resistant.

one-half-inch in diameter (16mm, as opposed to 18mm for "regular" one-half-inch solid drip hose) and comes with an emitter preinstalled inside the tubing at regular intervals. These internal emitters utilize what is known as a "tortuous path" [See Figure #43], in which the water must pass through a labyrinth of right-angled channels inside the emitter before exiting via a hole much larger than that of a typical punched-in emitter. The tortuous path causes the water to form a continuous vortex, a kind of horizontal tornado that keeps any water borne sediment, sand or silt in suspension so it won't settle out and clog the emitter.

In-line emitter tubing moistens the soil the entire length of the line and slightly below the surface, where the bulbous-shaped wet spots come together to form one nearly continuous moist zone. [See Figure #7, page 25 in the main text.] The emitters come pre-installed in tubing with 12-, 18-, 24-, and 36-inch spacings. The types of in-line tubing most commonly sold to gardeners are those with 12-inch or 24-inch intervals, but you can usually special-order tubing with emitters at 18-inch intervals. The emitters inside the hose are rated to dispense either one-half or one gallon per hour (actually .6 and .92 gph). The emitters in early versions of in-line tubing were non-pressure compensating, which meant that the flow rate at the end of the line might be lower than that near the main assembly. Newer versions include tubing with pressure-compensating emitters at the same intervals as in the non-compensating in-line tubing and with one-half- or one-gph flow rates.

I've used in-line emitters for over 20 years, and even with well water full of soluble iron oxide, (notorious for clogging regular punched-in emitters), I've found only a few clogged emitters in nearly a thousand feet of tubing. I always use the kind with one-half gph emitters on 12-inch centers because they will irrigate both sandy and clayey soils, depending upon how long the system is left on.

The benefits of in-line pressure-compensating emitters are many: they are very easy to install, suffer less clogging than porous tubing and some punched-in emitters, work at the greatest range of pressures (7–25 psi), provide consistent rates of irrigation on slopes totaling up to 10 feet high, and have no external parts to snap off. Their connectors or fittings don't leak. And the connectors, whether compression fittings or Spin Loc™ fittings, seal better than metal hose clamps with porous hose.

There are two main producers of in-line drip irrigation tubing in the USA. Both are in California, but market to distributors all over the country. They are: GEOFLOW™, which features Treflan®- impregnated emitters, [more on this later] and Netafim™. Both have pressure- and non-pressure-compensating emitters. Use the pressure-compensating ones if your total elevation change, both up and down, is greater than five feet or if you want to make sure your garden or lawn gets reliable, even distribution of water. I always use the pressure-compensating emitter tubing.

The single most important aspect of this type of watering system is the proper placement of the in-line drip tubing. Consistent irrigation of the entire root system is a must to ensure even, healthy turf or perennial plant growth and to prevent roots from invading the buried emitters. Two important horizontal criteria are the distance between emitters along the tubing and the measurement between the laterals (lengths of tubing). Usually, these two distances are the same. Generalized spacing guidelines vary considerably, depending on the tubing manufacturer or designer. For example, GEOFLOW recommends equal lateral and emitter spacings, with a 12- to 18-inch spacing in sandy soils, 18-inch intervals in clayey soils, and 24 inches for clay-loams. The apparent anomaly of the narrower interval for clayey soils is due to the inherent mechanical resistance of clay to horizontal water movement. In clay soils, the lines and emitters must be closer so that not too much water is lost downwards to gravity beneath the root zone before the water begins to bulge sideways.

At the Center for Irrigation Technology, at the California State University at Fresno, Field Research Manager Greg Jorgensen has installed GEOFLOW non-compensating one-half gph tubing on an 18-inch-lateral-distance-by-15-inch-emitter-interval on one acre of the University's high-traffic lawn. (GEOFLOW no longer makes the 15-inch interval tubing but sells it with 12-, 18- and 24-inch emitter intervals.)

Netafim's recommendations are 18-by-18-inch emitter spacings in clay, 12-inch-by-18-inch spacing in loam and 12-by-12-inch intervals in sand. Dennis Hansen, a landscape architect located in Sausalito, CA, has accumulated over 30 years experience with subsurface irrigation. He generally keeps his line and emitter spacings at 18 inches for lawns.

## The Tubing Layout

Since the soil texture ultimately determines these important measurements, you'll have to test your soil in various areas. To do so:

• Lay out some in-line emitter tubing next to the tubing in an area that seems to represent your yard's typical soil, connect it to a water source, and run water through it for a half hour.

• Dig a shallow trench to see how deep and wide some of the wet spots have spread beneath each of several emitters.

• Run the water for another half hour.

• Repeat digging the trench to follow the range of the wet spots. Shave away at the sides of the trench to see how wide the moist spot is getting.

This will give you an idea of the appropriate interval along the length of the tubing needed for emitters to maintain efficient irrigation, that is (as mentioned above), to create bulbous underground moist spots that converge beneath the surface of the soil.

## Tubing Depth

This is a project for a new lawn and can't be done with existing residential lawns. For lawns, the depth at which the emitter tubing is placed is a very important factor. Regardless of the choice of spacing between emitters, the tubing in the best lawn drip systems is laid out at a level six to ten inches below the soil surface. According to GEOFLOW, a six-inch depth is best for lawns, shrubs and trees in home landscapes. Dennis Hansen has found that four inches deep—plus or minus one inch—is best. NOTE: If you use a lawn aerator, be sure to set the tubing below the level of the aerator's piercing spikes.

## Not for Gophers!

If your garden is plagued by gophers, do not use buried irrigation as the pesty rodents will just chew through the tubing to get a free drink of water. However, in the over 20 years I've placed the tubing on the soil's surface and then heavily mulched the area, gophers have not eaten any of my tubing—cross my fingers! At the Center for Irrigation Technology, tubing was buried five inches deep (+/-) and suffered no gopher damage.

BUT, when they installed the SDI 16–18 inches deep in a new vineyard where alfalfa had been grown, they "had significant damage." Also, if squirrels live in your neighborhood, you might want to try a small test plot to see if they dig down to chew on the tubing.

## The Air Vacuum Relief Valve

Once your tubing is in place, there are, according to the manufacturers, two more very important gizmos to assist the emitters in remaining unclogged. An air vacuum relief valve is a must for a successful subsurface drip system [see Figure #70]. This simple and inexpensive part is placed at the highest point of the tubing

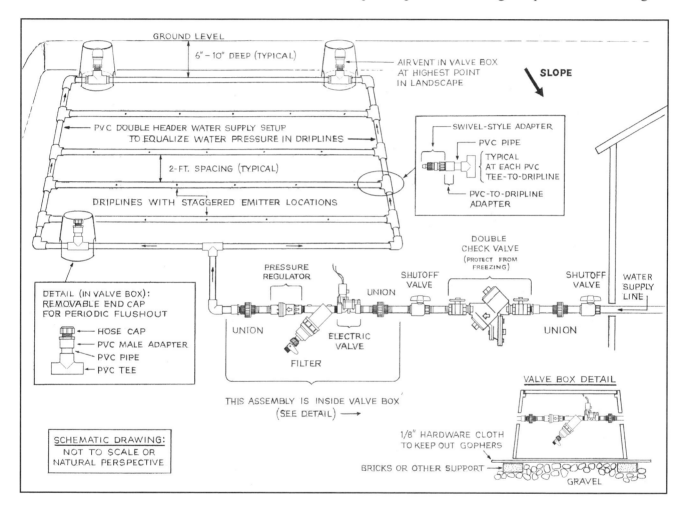

**Figure #70:** Subsurface drip irrigation demands careful placement of the in-line tubing—the distance between emitters along the laterals (lengths of tubing) and the measurement between the laterals—to ensure even irrigation. Soil texture determines these important measurements. Guidelines vary, from 24-inch-by-24-inch emitter spacing in clay, 12-inch emitter and 18-inch lateral spacing in loam and 12-inch-by-12-inch intervals in sand. A tubing depth of six- to 10-inches is the common recommendation. A double check valve is required when any of the tubing in the landscape is 12 inches or more above the source of the water so siphoning of the irrigation water doesn't contaminate the home's water. The double check valve must be installed only by a licensed contractor. (This is a schematic of an installation. The actual parts and specific adapters will vary with each installation. Read the in-line tubing manufacturer's on-line installation manual carefully or hire a landscape irrigation contractor .)

in each zone (also called a subassembly) attached to one valve. When the water comes on, the pressure quickly shuts this valve—it acts much like a check-valve. An air vacuum relief valve must be placed at every high point in an undulating or bermed landscape (a berm is a bank of earth often used as a retaining wall or hilly landscape feature). On the other hand, in talking with Dennis Hansen, he finds no need for this component as long as the system is on level ground and there's a flushing mechanism at the end of each subassembly. "The designers of air release valves," he cautions, "have not figured out how to keep out earwigs and other insects, which can clog the device." He notes that a new in-line product from Netafim™ has anti-backflow emitters built into the tubing, and is well adapted to slopes.

The importance of the air vacuum relief valve becomes evident when the system shuts off; as the pressure drops, the valve opens and air rushes into the tubing, helping to dry it out while also relieving the pressure on water draining out at the end of the line. The drier the inside of the tubing is between irrigations, the harder it is for searching root hairs to find a way inside, and the bloom of emitter-clogging algae is also reduced. The air vacuum relief valve should be 6–12 inches higher than the tubing, or at least 6 inches above the soil surface, whichever is higher. For its protection, this valve is usually placed inside a standard circular valve box.

Or, use a line flushing valve [as seen in Figure #8, page 27] at the lowest point of the system. This gizmo is also prone to failure do to insects and other critters. Check both the air vacuum relief valve and the line flushing valve often for "bugs" clogging the valve(s).

## Winterize

In really cold climates, it's safer at the end of the growing season to flush the above-ground main assembly [as seen in Figure #68, page139], drain it and store it in the garage or basement. Remove the line-flushing cap and drain the drip hosing. There are also fittings available that allow small air compressors to purge the drip hose before winter. Or, simply open the valve at the lowest point of the system to drain out any remaining water. (This same valve is used to flush the lines in the spring and as needed through the growing season.)

## How Much Water?

It's nice to know, according to literature provided by Toro Company (the manufacturer and distributor of lawn mowers, lawn equipment, and irrigation products) that SDI systems make a "46% larger wetted volume of soil than a surface drip system. This decreases the saturation point of the soil, which not only leaves room for more air…[but] decreases the water lost to deep percolation."

The best way to time the watering of your lawn is to use a chart based upon your climate's evapotranspiration rate, instead of guessing or using an arbitrary rule of thumb. Your local Cooperative Extension office or Water District's Water Conservation Department should be able to tell you each month's average ET rate in inches. The chart entitled Daily Water Use in Figure #46 [page 81] converts the monthly ET rate from inches-per-month into gallons-per-day for one-square foot of lawn or garden space up to one full acre.

Watering to equal the ET rate is a good starting point. In a well-drained soil, you can, if desired, apply more water than the suggested ET rate to encourage more growth. Or, if your water supply is limited, back off from the amount suggested by the ET rate. As an example, in the coastal zone of Northern California near San Francisco, where I live, lavender can do well with only 25% of the water listed on the chart. (Actually, I plant lavenders in the fall and never irrigate them again and they do fine.)

## Timing is Everything

Subsurface lawn systems require short, frequent daily irrigations. One irrigation cycle per day is the minimum interval at the Center for Irrigation Technology, but multiple start times are preferred. Landscaper Dennis Hansen suggests "unlimited very brief start times, actually up to 40 times per day, run by a controller with independent stations (programs) to have control over separate zones.

GEOFLOW™ recommends a single daily watering, but, because it's impregnated with a substance they call Rootguard (the chemical Treflan®; see the following footnote), every-other-day irrigations will work, with no root intrusion.

The duration of each application is calculated by dividing by the number of start times per day into the total length of irrigation needed each day. The length of each of Hansen's irrigation cycles, however, is based upon achieving "no dry zones beneath the surface; I run the drip system until there is a solid, thin blanket of water across the surface." No long run times are required because the feeding root hairs of lawns, trees and shrubs occupy only the upper 12 to 24 inches of soil.

## Treflan®, *Not* for Organic Gardeners

Treflan® is a relatively benign, but by no means "organic," chemical that's been around for 30-

plus years and acts as a growth inhibitor by arresting cell division (it stops the apical root bud from dividing). Applied in a granular form, Treflan® can control the sprouting of annual seeds. According to DowElanco, Treflan®, doesn't dissolve in water and doesn't leach into the soil because it adheres to the clay.

FOOTNOTE

Studies by the Center for Irrigation Technology confirm the effectiveness of the root-excluding properties of GEOFLOW's emitters imbedded with Treflan®. According to CIT Field Research Manager Greg Jorgensen, "In 1990, after three years of use with tall fescue grass, there was a root-free sphere the size of a golf ball around each emitter." Later studies found large, malformed roots alongside the emitter, but no intrusion. All this would seem to indicate that the Treflan does, in fact dissolve a bit. It does not, however, translocate into plant stems, leaves, fruits, or flowers.

The compound has a low $LD_{50}$ (a controversial way to rank chemical toxicity) of 10,000 mg/kg—the smaller the number, the more toxic the compound; the $LD_{50}$ of very strong organic pesticide such as nicotine sulfate is 50-91 mg/kg. Testing in 1977–78 showed Treflan-related patacellular carcinoma (cancer) in female mice, but the Treflan formula used was contaminated by nitrosamines, which are now kept, according to DowElanco, "well below 0.5 ppm." According to a DowElanco spokesperson, out of eight subsequent tests, only one, at the relatively high rate of >4000 ppm dermal exposure for rats, "showed a possible cancer potential for people."

FOOTNOTE

## In-Line Drip Hose Resources:

You may have to look around for a supplier of in-line tubing. Check in the Yellow Pages under Irrigation Systems and Equipment. Or, contact the following manufacturers:

The Techline™ in-line emitter tubing (pressure compensating, *without* Treflan®, my personal choice) is distributed to the trade by: Netafim™, 3025 East Hamilton, Fresno, CA 93721; 1-(800) 777-6541. Call to find out the distributor closest to you, who will probably tell you of local retail outlets. Or, check out their Web site: http://netafimusa.com. Click "Landscape & Turf" under the header "Divisions." Scroll down to the "Support & Services" box and click on "Where to Buy." Type in your zip code, and you'll get a list of local retail and wholesale suppliers. (www.netafim.com is the international site for access to Netafim™ products anywhere in the world.)

GEOFLOW™ Dripline with Rootguard® (pressure compensating, with Treflan®) is distributed to the trade by: GEOFLOW, 200 Gate 5 Road, #103, Sausalito, CA 94966; 1-(800) 828-3388, or (415) 927-6000, (http://geoflow.com) Call to find out about distributors, and, if you're nervous about installing the tubing yourself, how to contact SDI irrigation contractors in your area. There also is a technical PDF download about design, installation, and maintenance

Toro Company (http://www.toroag.com ) sells their pressure-compensating in-line tubing with Rootguard® (Treflan®) to landscapers as "DL2000". They also have a line called Drip In® Classic (with Rootguard®, Treflan®), which is marketed to farmers but will work for homeowners, Toro Company doesn't offer tubing with a 12-inch emitter spacing, only 18 inches and up. You can get a detailed design "manual" for SDI by going to http://www.toroag.com and print the PDF file.

# Appendix #2

# Legumes to Improve Your Soil

If you have less than desirable soil for a no-till garden, green manuring for a number of years can really improve the soil. Green manuring is a term used to subscribe the practice of digging or turning fresh green plants into the soil to improve its quality, fertility, and structure, a time-honored cultivating "tool." If your soil is too clayey, green manures can improve it, both by the soil-loosening action of the living plants' roots and then by the addition of the tilled-in foliage. Green manures can also add fiber and nutrients to a sandy soil, increasing its ability to hold onto moisture and nutrients.

If you choose to till or dig your garden in your usual way, planting some of the garden with a green manure crop—usually composed of a mixture of grasses, grains, and legumes—will help provide nitrogen, phosphorus, and many micronutrients to the soil. Rotate where you plant the green manure crop so the entire garden will get the advantage of the nutrients and the additional fiber—sort of like a buried compost pile. However you use green manure, it's close to free fertilizer—not counting the cost of a few seeds and plenty of elbow grease.

If the plant matter is turned into the soil when the plants are fresh, green, and high in nitrogen, then the soil bacteria can easily and readily decompose the plant tissues. One of the premier virtues of green manures is that the plant matter decomposes so quickly that you can plant soon after tilling it into the soil. Usually, the prudent gardener need only wait two to four weeks in warm weather after tilling under a green manure before transplanting or seeding.

At least once each spring, plant a cover crop—which is to become a green manure—and wait until it is about four- to six-inches high. Then, till it under with a garden fork or rototill. Or better yet, sow and till under several crops of green manures in one season to really give your garden's soil a boost. You can use the list below to choose a crop to plant in the fall for tilling under in the spring—although the tops may have to be cut in order to have only enough stubble to make the crop easy to till under. (The tops go off to the compost pile or an area already under no-till practices.)

The optimal time to till under a green manure crop of nitrogen-fixing legumes (for a subsequent crop) is just before the blooms appear. The conscientious green manurer must "sacrifice" beautiful blossoms for the optimal amount of nitrogen.

I was visiting a friend one year when almost one third of their front yard had crimson clover in full, glorious hot-red bloom. I timidly mentioned that the best effect for nitrogen occurs prior to bloom, or up to 20 percent of bloom. "Yes," he replied, "I won't get as much nitrogen, but look how absolutely gorgeous the bloom is! I don't mind giving up a bit of the nitrogen for the floral display. Besides, Crimson Clover has a decent amount of nitrogen to spare, and my soil isn't in horrible shape." Good choice.

If you already have a healthy crop of leguminous plants or vegetables, you can skip the addition of any type of phosphorus. Once legumes begin to grow, they accumulate plenty of phosphorus themselves and don't require supplemental phosphorus fertilizer. If you keep tilling your legumes under, the phosphorus content of the soil will also increase.

Most legumes prefer not to grow in an acid soil. Take a soil test and adjust the pH to 6.5 to 7.

Most legume seeds will thrive even more if they are inoculated with the correct rhizobium bacteria. Ask your supplier what inoculant is best for the legume(s) you have chosen. Some seeds come already coated with a thin layer of inoculant.

Here are some legumes to consider:

- Alfalfa: Has very deep roots—six or more feet deep—to gather micronutrients. Don't let this perennial legume get established as it can become a perennial "weed;" till it under before bloom so it can't set seed. If you want to keep this perennial, it's best left in pathways and mowed on a regular basis. Watch for spreading roots, and cut the roots off between the pathways and all planted areas with a straight-back garden spade on a regular basis.

- Beans: Mung, soy, and velvet. Good for warm soils. Plant late spring or early summer.

- Bur Clover: A good plant to seed in the fall in warm winter climates. One of the best legumes for mild climates.

- Crimson Clover: Will grow well in slightly acid soils. Not winter hardy north of New Jersey. Can be planted in the fall elsewhere.

- Cow Pea: Withstands drought and will tolerate some shade. Must be planted after the soils have warmed up.

- Field Peas: Spring sowing in cold climates, fall planting in warm winter zones.

- Lupines: Thrive on acid soil low in fertility.

- White Clover: Very hardy. Fall plantings winter over. Don't let this one go to seed as it will become invasive. Like alfalfa, it's a perennial and spreads. Best left in pathways and mowed on a regular basis. As with alfalfa, cut any spreading roots off with a straight-back garden spade.

- Vetch, Hairy: Not winter hardy in severe climates. Till under before seed forms or it will become a "pest." It twines around anything it can find, including trellises and fences.

# Appendix #3
# Searching for Tube-Grown Plants

Tube-grown plants are very hard to find from mail-order companies. One of the best local resources may be nurseries that grow trees and shrubs for revegetation programs or habitat restoration. Ask your local native plant society for any leads for revegetation nurseries. (Of course, you'll only get native plants; however, they are the ones best suited to your microclimates.) In some areas, the forest service sells small seedlings for replanting logged forests and to help homeowners establish a native forest. Here are a few resources. If you can't find tube stock locally or through the mail, go for 4" pots whose roots aren't "pot-bound."

**Digging Dog Nursery**
P.O. Box 471
Albion, CA 95410
Phone (707) 937-1130
Fax (707) 937-2480
www.diggingdog.com
They carry a wide selection of hard-to-find perennials. But, they also have some unusual trees—such as the Dove Tree (*Davidia involucrata*). Catalog is $4.

**Forest Farm**, Ray and Peg Prag
990 Tetherow Rd.
Williams, OR 97544-9599
(541) 846-9230
From their Web site (www.forestfarm.com) "Tubes for most woody plants are as deep as a gallon can (6") but approx. 2"x 2" wide, making them much lighter (about less than a pound) to ship. Tubes for some perennials and many woodland trees (if they're surface-rooted) are only 4–5" deep and some tubes are plastic, rather than paper. The plant tops can vary between 3" and 3', but are generally 6–18"." Their catalog

is 500 pages and lists thousands of plants; some common, but mostly unique or rare. Tube-grown plants are available for many of the selections. Catalog is free.

**Greer Gardens**
1280 Goodpasture Island Road
Eugene, OR 97401
1-(800)-548-0111
www.greergardens.com
In an e-mail from the nursery: "While most of the plants we sell are gallon-size or larger, we also supply some plants in small containers, mostly 4". We ship very little that is true bare-root, though we do take most of the plants out of containers and put them in plastic bags to save weight and space."

**Itasca Greenhouse**
P.O. Box 273
Cohasset, MN 55721
1-(800)-538-TREE
(218) 328-6261
igtrees@northernnet.com
Sells both native and exotic species of trees and shrubs by mail in 2.3 cubic in. and 3.7 cubic in. soil "plugs" (like tubes) and various other small containers. Prices vary according to volume. Also lists a good set of supplies to assist your purchases' growth, including mycorrhizal inoculants. Catalog is free.

**Plants of the Wild**
P.O. Box 866
Tekoa, WA 99033
Phone (509) 284-2848
Fax (509) 284-6464
plants@eznet.com

www.plantsofthewild.com

Offers a wide selection of trees and shrubs native to the country in 4 cubic in. or 10 cubic in. "plugs" (like tubes) and 3 in. pots. Also lists wildflowers and bunchgrasses. Catalog is free.

# Appendix #4
# A Short List of Shrubs

Those marked * can also develop into tree form. Most have more than one species and/or variety. So I skipped using the "spp." so it won't be boring.

- *Abelia* (Abelia)
- *Arctostaphylos* (Manzanita) *
- *Aronia* (Chokeberry)
- *Artemisia* (Sagebrush)
- *Berberis* (Barberry)
- *Buddleja* (Butterfly Bush)
- *Buxus* (Box) *
- *Callistemon* (Bottlebrush) *
- *Calluna* (Heather)
- *Calycanthus* (Spice bush)
- *Camellia* (Camellia) *
- *Caragana* (Pea tree) *
- *Carpenteria* (Carpenteria)
- *Caryopteris* (Bluebird)
- *Ceanothus* (Ceanothus) *
- *Ceratostigma* (Plumbago)
- *Cercocarpus* (Mountain mahogany) *
- *Chaenomeles* (Japanese quince)
- *Chionanthus* (Fringe tree) *
- *Choisya* (Mexican-orange Blossom) *
- *Cistus* (Rockrose)
- *Clethra* (Summersweet, Sweet Pepperbush) *
- *Cornus* (Dogwood) *
- *Corylopsis* (Winter hazel) *
- *Cotinus* (Smoketree) *
- *Cotoneaster* (Cotoneaster) *
- *Crataegus* (Hawthorn) *
- *Crinodendron* (Crinodendron) *
- *Daboecia* (Irish heath)
- *Daphne* (Daphne)
- *Dendromecon* (Bush Poppy)
- *Deutzia* (Deutzia)
- *Elaeagnus* (Elaeagnus) *
- *Embothrium* (Chilean fire bush) *
- *Ephedra* (Ephedra)
- *Erica* (Heath)
- *Eriobotrya* (Loquat) *
- *Escallonia* (Escallonia)
- *Eucryphia* (Eucryphia) *
- *Euonymus* (Spindle) *
- *Forsythia* (Forsythia)
- *Franklinia* (Franklinia) *
- *Fremontodendron* (Flannel bush)
- *Fuchsia* (Fuchsia) *
- *Garrya* (Silk-tassel) *
- *Gaultheria* (Salal)
- *Grevillea* (Grevillea)*
- *Hamamelis* (Witch hazel) *
- *Hebe* (Hebe)
- *Helianthemum* (Rockrose)
- *Hibiscus* (Hibiscus) *
- *Hippophae* (Sea buckthorn) *
- *Hoheria* (Lacebark) *
- *Hydrangea* (Hydrangea)
- *Hypericum* (Rose of Sharon)
- *Ilex* (Holly) *
- *Jasminum* (Jasmine)
- *Juniperus* (Juniper) *
- *Kalmia* (Mountain laurel)
- *Kolkwitzia* (Beautybush)
- *Lagerstroemia* (Crape myrtle) *
- *Lavandula* (Lavender) a sub-shrub
- *Lavatera* (Tree Mallow)
- *Lespedeza* (Bush Clover) *
- *Leptospermum* (Manuka) *
- *Ligustrum* (Privet) *
- *Lindera* (Spicebush) *
- *Linnaea* (Twinflower)
- *Lupinus* ( Lupine)

- *Lycium* (Boxthorn)
- *Magnolia* (Magnolia)*
- *Mahonia* (Mahonia)
- *Myrica* (Bayberry) *
- *Myricaria* (Myricaria)
- *Myrtus* and allied genera (Myrtle) *
- *Osmanthus* (Osmanthus)
- *Pachysandra* (Pachysandra)
- *Perovskia* (Russian Sage)
- *Philadelphus* (Mock-orange) *
- *Phlomis* (Jerusalem Sage)
- *Photinia* (Photinia)
- *Pieris* (Pieris)
- *Pittosporum* (Pittosporum)
- *Potentilla* (Cinquefoil)
- *Pyracantha* (Firethorn)
- *Rhamnus* (Buckthorn) *
- *Rhododendron* (Rhododendron, Azalea) *
- *Ribes* (Currant)
- *Romneya* (Tree poppy)
- *Rosa* (Rose)
- *Rosmarinus* (Rosemary)
- *Rubus* (Bramble)
- *Sambucus* (Elderberry) *
- *Santolina* (Lavender cotton)
- *Senecio* (Senecio)
- *Sophora* (Mescalbean) *
- *Spiraea* (Spiraea) *
- *Symphoricarpos* (Snowberry)
- *Syringa* (Lilac) *
- *Vaccinium* (Bilberry,huckleberry, foxberry)
- *Viburnum* (Viburnum) *

# Appendix #5

# More Trees That Can Also Grow in Lawns

| **Common Name** | *Latin Name* |
|---|---|
| Trident Maple | *Acer buergeranum* |
| Box Elder | *Acer negundo* |
| Red Buckeye | *Aesculus pavia* |
| Australian Flame Tree | *Brachychiton acerifolius* |
| Common Hackberry | *Celtis occidentalis* |
| Eastern Redbud | *Cercis canadensis* |
| Deodar Cedar | *Cedrus deodara* |
| Brush Cherry | *Eugenia* spp. |
| Ash | *Fraxinus americana* 'Autumn' Purple' (seedless) |
| Gingko, Maiden Hair Tree | *Ginkgo bilboa* |
| Goldenrain Tree | *Koelreuteria paniculata* |
| Crape Myrtle (one of the new mildew-resistant cultivars) | *Lagerstroemia* spp. |
| Crab Apple | *Malus* x 'Robinson' |
| Flowering Crab Apple | *Malus floribunda* |
| Ornamental/Chinese Pistache | *Pistachia chinensis* |
| Fern Pine | *Podocarpus gracilior* |
| Flowering Cherry | *Prunus* 'Okame' |
| English Laurel | *Prunus laurocerasus* |

| | |
|---|---|
| Holly Oak | *Quercus ilex* |
| Red Oak, Northern Red Oak | *Quercus rubra* |
| Chinese Scholar or Japanese Pagoda Tree | *Sophora japonica* |
| Oriental Arborvitae | *Thuja orientalis* |

# Appendix #6

## Some Trees and Shrubs Susceptible to
## *Phytophthora cinnamoni* and *P. lateralis*

Reprinted with permission of The Scotts Company, LLC

*Abies* spp. – Fir.

*Acacia* spp. – Acacia

*Arctostaphylos* spp. – Manzanita

*Calluna vulgaris* – Scotch heather

*Camellia japonica* – Camellia

*Castanea* spp. – Chestnut

*Casuarina* spp. – Beefwood, She oak

*Ceanothus* spp. – Wild lilac, Tick brush

*Cedrus* spp. – Cedar

*Chamaecpyaris* spp. – False cypress

*Daphne* spp. – Daphne

*Erica* spp. – Heath

*Eucalyptus* spp. – Eucalyptus

*Fastia japonica* – Japanese aralia

*Hibiscus* spp. – Hibiscus

*Hypericum* spp. – St. John's Wort

*Juglans* spp. – Walnut

*Juniperus* spp. – Juniper

*Larix* spp.– Larch

*Laurus noblis* – Sweet bay

*Myrtus communis* – Myrtle

*Olea europea* – Olive

*Picea* spp. – Spruce

*Pieris* spp. – Andromeda

*Pinus* spp. – Pine

*Pittosporum* spp. – Mock orange

*Platanus* spp. – Sycamore

*Pseudotsuga menziesii* – Douglas fir

*Quercus* spp. – Oak

*Rhododendron* spp. – Rhododendron, Azalea

*Salix* spp. – Willow

*Sequoia sempervirens* – Coastal redwood

*Taxodium distichum* – Bald cypress

*Taxus* spp. – Yew

*Thuja* spp. – Arborvitae

*Viburnum* spp. – Viburnum

# Fruit Trees Susceptible to *Phytophthora cinnamoni* and *P. lateralis*

Apricot

Avocado

Blueberry, Highbush (Rabbiteye is resistant)

Cherry

Citrus

Peach

Pear

## Resistant Trees and Shrubs:

*Camellia sasanqua* – Camellia

*Chamaecyparis nootkatensis* – Alaska cypress

*Chamaecyparis pisifera* var. *filifera* – Sawara

*Chamaecyparis thyoides* – White cedar

*Daphne cneorum* – Rock daphne

*Juniperus chinensis* 'Pfitzerana' – Pftizer's juniper

*Juniperus sabina* – Savin juniper

*Juniperus squamata* 'Meyeri' – Meyer juniper

*Pinus mugo* var. *mugo* – Mugo pine

*Rhododendron obtusum* – Hiryus azalea

*Thuja occidentalis* – Arborvitea

# ❧ Bibliography ☙

Alexander, Martin. *Introduction to Soil Microbiology*. New York: John Wiley & Sons, 1961.

Alth, Max. *How to Farm Your Backyard the Mulch-Organic Way*. New York: McGraw-Hill, 1977.

Bahrt, George M., et al. *Hunger Signs in Crops*. Ed. Gove Hambidge, Washington, DC: The American Society of Agronomy and The National Fertilizer Association, 1941.

Brooks, J.R., et al. **"Hydraulic Redistribution of Soil Water During Summer Drought in Two Contrasting Pacific Northwest Coniferous Forests."** *Tree Physiology* 22 (2002): 1107-111.

Buckman, Harry O., and Nyle C. Brady. *The Nature and Properties of Soils*. New York: The Macmillan Company, 1963.

Burgess, Stephen O., et al. **"Seasonal Water Aquisition and Redistribution in the Australian Woody Phreatophyte, *Banksia prionotes*."** *Annals of Botany* 85 (2000): 215-224.

Caldwell, M. Martin, Todd E. Dawson, K. James, and H. Richards. **"Hydraulic Lift: Consequences of Water Efflux from the Roots of Plants."** *Oecologia* 113 (1998):151-161.

Easy, Ben. *Mother Earth*. (UK) January 1952 and October 1954.

Eis, S. **"Natural Root Forms of Western Conifers."** Rpt. in *Proceedings of the Root Form of Planted Trees Symposium*, Departmental Report No. 8., 23-27, Victoria, B. C.: British Columbia Ministry of Forests/Canadian Forest Service, 1978.

Espeleta, J. F., J. B. West, and L. A. Donovan. **"Species-specific Patterns of Hydraulic Lift in Co-occurring Adult Trees and Grasses in a Sandhill Community."** *Oecologia* 138 (2004): 341-349.

Foxx, Teralene S., Gail D.Tierney, and Joel M.Williams. **"Rooting Depths of Plants on Low-Level Waste Disposal Sites."** Produced for the Los Alamos National Laboratory Report #LA-10253-MS. November 1984.

Foxx, Teralene S., and Gail D. Tierney. **"Root Lengths of Plants on Los Alamos National Laboratory Land."** Los Alamos National Library, #LA-10865-M. Jan. 1987.

Harris, Richard W., James R. Clark, and Nelda P. Matheny. *Arboriculture-Integrated Management of Landscape Trees, Shrubs and Vines*. Upper Saddle River, NJ: Prentice-Hall, 1999.

Jeavons, John. *How to Grow More Vegetables - Than You Ever Thought Possible on Less Land Than You Can Imagine.* Ecology Action, Willits, CA: Ten Speed Press. 2007.

Kolsnikov, V. A. *The Root System of Fruit Trees.2nd ed.,* The U Press of the Pacific, 2003.

Koo, B-J., and Adriano, D. C. *Root Exudates and Microorganisms*. U of Georgia, Aiken, SC; Massey U, Palmerston North, New Zealand, Barton C.D., U of Kentucky, Lexington, KY. *Encyclopedia of Soils in the Environment,* D. Hillel ed., Amsterdam: Elsevier Academic Press, 2005. 421–428.

Kourik, Robert. *Designing and Maintaining Your Edible Landscape—Naturally*. Occidental, CA: Metamorphic Press, 1986. (Reprinted: East Meon, Hampshire, UK: Permanent Publications. 2004)

- - - . *Drip Irrigation, For Every Landscape and All Climates*. Occidental, CA: Metamorphic Press, 1992.

- - - . *The Lavender Garden*. San Francisco: Chronicle Books, 1998.

- - - . *Pruning, clipping with confidence*. New York: Workman Publishing Company, 1997.

- - - . *The Tree & Shrub Finder*. Newtown, CT: Taunton Press, 2000.

Krasilnikovof, N. A. *Soil Microorganisms and Higher Plants*. N. A. Academy of Sciences of the USSR, Institute of Microbiology. Moscow: Academy of Sciences of the USSR, Moscow, 1958. Reprinted for The National Science Foundation, Washington, DC and The Department of Agriculture, by The Israel Program for Scientific Translations, 1961.

Linden, Dennis. **"ARS Quarterly Report, Oct. 1 to Dec. 31"**, USDA Agricultural Research Station, 1997. St. Paul, MN, Linden, Dennis. Telephone interview. May 21, 2006.

McCulley, R. L., et al. **"Nutrient Uptake as a Contributing Explanation for Deep Rooting in Arid and Semi-arid Ecosystems."** Oecologia. 141 (2004): 620-628.

O'Brien, Kenneth Dalziel. *Veganic Gardening, the alternative system for healthier crops*. Rochester, VT: Thorsons Publishing Group, 1986.

O'Brien, Rosa Dalziel. *Intensive Gardening, Using Dutch Lights, Surface Cultivation and Composting for the Commercial Production of Crops, and Introducing a Motion-Study Routine*. London, UK: Faber & Faber Limited, 1956.

Perry, Thomas O., ed. *The Landscape Below Ground*. Savoy, IL: International Society of Arboriculture, 1994.

Quinn, Vernon. *Roots, Their Place in Life and Legend*. New York, NY: Frederique A. Stokes Company, 1938.

Russell, E. W. *Soils Conditions and Plant Growth*. New York: Longman Group, Inc., 1973

Solomon, Steve. *Water-Wise Vegetables*. Cascadia Gardening Series. Seattle: Sasquatch Books, 1993.

Stout, Ruth. *How to Have a Green Thumb Without an Aching Back*. New York: Simon & Schuster, 1955.

- - - . *Gardening Without Work; for the Aging, the Busy and the Indolent*. New York: Devin-Adair, 1969.

Stout, Ruth, and Richard Clemence. *The Ruth Stout No-Work Garden Book*. Emmaus, PA: Rodale Press, 1971.

Sylvia, D.M., et al., eds. "**Mycorrhizal Symbioses,**" *Principles and Applications of Soil Microbiology*. 2nd ed. Upper Saddle River, NJ: Prentice-Hall, 2005. 263-282.

Sylvia, David M. *Principles and Applications of Soil Microbiology: 2nd Edition*, Upper Saddle River, NJ: Prentice-Hall, 1998. (Box 12-4, 278).

Tinus, Richard, and Stephen McDonald. *How to Grow Tree Seedlings in Containers in Greenhouses*. Fort Collins, CO: Rocky Mountain Forest and Range Experiment Station, Forest Service; U.S. Dept. of Agriculture, 1979.

Waisel, Yoav. *Plant Roots: The Hidden Half*. Eds. Amram Esheland and Uzi Kafkafi. New York: Marcel Dekker, Inc. 2002.

Watson, Dr. Gary, and Dr. Dan Neely, eds. **"The Landscape Below the Ground"**. *Proceedings of International Workshop on Tree Root Development in Urban Soils*. Savoy, Il: International Society of Arborists, 1993.

Weaver, John, and William Bruner. *Root Development of Vegetable Crop*s. New York: McGraw Hill Book Company, 1927.

Weaver, John, and Frederic Clements. *Plant Ecology*. New York: McGraw-Hill Book Company, 1938.

Whitcomb, Carl E. *Landscape Plant Prodution, Estabishment, and Maintenance*. Stillwater, OK: Lacebark Publications, 1986.

Note: Page numbers in *italic* indicate an illustration.

# Other Useful Gardening Books by Robert Kourik

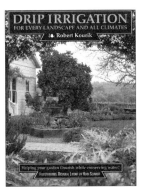

**118 Pages, Softcover**
**$15 USD**

### Drip Irrigation for Every Landscape and All Climates

provides step-by-step instructions for assembling custom-designed drip systems for every part of your landscape. It also details and describes the highest-quality parts for each project, and lists reliable suppliers from whom they're available. (Forget about those pre-packaged kits unless you want to wind up with one or more bogus parts.)

This remains the only detailed book written exclusively about drip irrigation for the home gardener. Lynne Ocone of *Sunset Magazine* called it "The last word on drip irrigation," and adds "[This] book is infused with good humor, abundant illustrations, and clear and logical explanations."

In 1996, the well-respected gardening newsletter the *Avant Gardener* selected *Drip Irrigation for Every Landscape and All Climates* as number seven on its list of "25 Best Books of the Past 25 Years."

**192 Pages, Hardcover**
**$25 USD**

### The Tree & Shrub Finder: Choosing the Best Plants for Your Yard

is a clearly written and practical guide designed to take the uncertainty out of selecting ornamental trees and shrubs for your home and garden. The book uses down-to-earth, clear, jargon-free, and up-to-date information to dispel common myths and misconceptions about buying, planting and caring for trees and shrubs. In addition, it provides a fresh and modern look at choosing yard plants for appearance and appropriateness

The book contains references and detailed listings for over 200 shrubs and trees. Twenty-one Plant Finder charts give for each plant: its common and Latin name, size, USDA zones, and a paragraph on the potentials and limitations of each plant.

Each Plant Finder chart can also help you choose trees and shrubs for: resisting hungry deer, attracting birds, wind tolerance, ability to withstand droughts, tolerance of heavy clay soils, and the ability to thrive in sandy soils.

**120 Pages, Hardcover**
**$16 USD**

### *The Lavender Garden*

Describes fifteen varieties of lavender in detail: common name, botanical name, flower description, bloom period, plant and foliage description, hardiness, planting range, and typical landscape use. ***The Lavender Garden*** also includes all you'll need to know about planting and caring for your lavender. As well as reviewing various crafts made from lavender, the book includes 13 recipes from entrees to breads and desserts.

The book was an instant success, mentioned in *Martha Stewart Living*, *House and Garden Magazine* and numerous other magazines and daily newspapers. It is still all the rage since it was first published.

"*The Lavender Garden*" is quickly becoming the bible of American lavender growers, the first book…published in America by an American for American lavender growers…His 120-page volume cuts through the arcane botanical nomenclature to identify and select 15 commonly-found favorite lavenders." September 12, 1998, *The Press Democrat*. Santa Rosa, CA

**371 Pages, Softcover**
**$40 USD**

### Designing and Maintaining Your Edible Landscape—Naturally

Gardening fads come and go, but national and international interest in a beautiful and tasty "edible landscape" continues unabated. One of the most respected texts on the subject is *Designing and Maintaining Your Edible Landscape—Naturally*, by Robert Kourik (Metamorphic Press). Originally published in 1986, the book continues to sell steadily as a nationally-favored gardening reference. It remains a classic text for gardeners of all skill levels. Now, after more than 20 years, this classic book is back in print.

Today, as in 1986, the **"Golden Rules of Edible Landscaping"** still apply:

- Start ever so small.
- Enjoy your landscape; if it's just drudgery, you're doing something wrong.
- Plant your vegetables no further from the kitchen sink than you can throw the kitchen sink.
- Be lazy; let nature work for you.
- Turn limitations into virtues.
- Time and money spent early means time and money saved later.
- Plan in advance; make your mistakes on paper, not in your landscape.
- Plan for the unexpected; nature will be unpredictable.
- Try to use plants that serve more than on use.

Dig in! Eat well!

---

You may photocopy this ORDER FORM for your convenience in ordering.

 Book prices include tax, shipping, and handling

Also available online at
www.robert-kourik.com

| QUANTITY | | PRICE EACH | TOTAL PRICE |
|---|---|---|---|
| | Edible Landscape | $40.00 | |
| | Tree & Shrub Finder | 25.00 | |
| | Lavender Garden | 16.00 | |
| | Drip Irrigation | 15.00 | |
| | (USD) TOTAL $ | | |

**Please send books to:**

NAME _____ STREET ADDRESS_____

CITY _____ STATE _____ ZIP _____

Please make your check payable to **Metamorphic Press** and mail your order to:

Metamorphic Press
P.O. Box 412
Occidental, CA 95465
USA

*Each book will be autographed by Robert Kourik*

♦ **Roots Demystified pulls together practical science from a wide range of sources**, including a unique assembly of detailed scale drawings of a variety of veggie and tree root systems. Even a seasoned and well-read plant person will find interesting and practical information in this well-written manual. For example, in my 27-plus years as a tree nurseryman, I had not learned that trees redistribute water in the soil profile via "hydraulic lift" [Chapter 11]! Soil fertility and irrigation are covered as well, made eminently useful presented in the context of where and how roots grow. —**Robert Woolley**, President, Dave Wilson Nursery, Hickman (near Modesto), CA

♦ Gardeners interested in sustainability will find *Roots Demystified* **an invaluable resource.** If roots are the key to sustainable gardening, Robert Kourik's Roots Demystified offers the key to growing terrific roots. The astonishingly detailed pictures are an education in themselves, earning this excellent new handbook a place on every gardener's reference shelf. —**Ann Lovejoy**, Author/Lecturer, Bainbridge Island, WA

♦ **A really great book!** Robert has taken a complex subject and refined it into a highly informative, must-read book for every ecologically-minded gardener. —**Chip Tynan**, Manager, Horticulture Answer Service, Missouri Botanical Garden, St. Louis, MO

♦ I saw my first root drawing in 1974, when I took a soils class while a student at the University of California at Berkeley. But, it didn't really hit home with me until looking at a root drawing in 1982 while reading John Jeavons' book, *How To Grow More Vegetables.* Robert has done an excellent job compiling all the research that has been done in the past twenty years to flesh out our knowledge of what plant roots do and how they grow. He has made very complicated information easily accessible and given practical information for gardeners to use in growing better plants. —**Kate Burroughs**, Founding Co-owner, Harmony Farm Supply, Sebastopol, CA

♦ Kourik's book about the significance of roots comes from many years of physical labor in the fields, not in a classroom. Join that demanding and exhausting physical labor to his ability to combine observation and analysis and words, and you have **a book more valuable to gardeners and diners and readers than any such book on the market**, past or current. And the illustrations? As my teenage great grandson says: "I mean, like, they're mind-blowing, man!" —**Chester Aaron**, Garlic Farmer/Author, Occidental, CA

♦ I thought Robert Kourik had written the most useful book on ecological gardening when he wrote, *...Your Edible Landscape—Naturally.* Apparently, he had only scratched the surface. Now he's written a new book, *Roots Demystified, Change Your Gardening Habits to Help Roots Thrive.* Robert has a rare gift for rendering complex scientific studies into a compost of humorous, fascinating environmental wisdom and easy-to-apply, in-the-dirt, hands-on knowledge. **After you read *Roots Demystified*, every tree you plant will dig in and thrive for you.** What Alex Haley taught us about how to value and nurture our metaphorical family tree, Robert has done for our actual family tree—the one in our backyard. —**Sam Benowitz**, Owner, Raintree Nursery, Morton, WA

♦ **Roots is an amazing look into the underground world of plants.** After showing us a world that we knew little about Robert gives us practical ways to apply this knowledge to our daily horticultural activities. —**Tom Bressan**, Owner, The Urban Farmer Store, San Francisco, CA

♦ Ever wonder why your prize perennial mopes and dwindles, or that newly planted tree looks the same size years later? It's the roots! Shedding light on a dark subject, this book illuminates a long out-of-sight topic with scientific facts and superb illustrations. **Gardeners should make room on their library shelves for *Roots Demystified*.** —**Mimi Luebbermann**, Author, Garden Writer, Petaluma, CA